Advances in
Image Pickup and Display

VOLUME 2

CONTRIBUTORS TO THIS VOLUME

J. J. Brandinger

D. B. Fraser

G. L. Fredendall

J. R. Maldonado

A. H. Meitzler

D. H. Pritchard

Teiichi Taneda

Manabu Yamamoto

Advances in
IMAGE PICKUP AND
DISPLAY

VOLUME 2

Edited by *B. Kazan*

XEROX CORPORATION
PALO ALTO, CALIFORNIA

ACADEMIC PRESS   New York   San Francisco   London   1975

A Subsidiary of Harcourt Brace Jovanovich, Publishers

ACADEMIC PRESS, INC.
111 Fifth Avenue, New York, New York 10003

United Kingdom Edition published by
ACADEMIC PRESS, INC. (LONDON) LTD.
24/28 Oval Road, London NW1

LIBRARY OF CONGRESS CATALOG CARD NUMBER: 73-18958

ISBN 0–12–022102–0

PRINTED IN THE UNITED STATES OF AMERICA

Contents

Laser Displays

Manabu Yamamoto and Teiichi Taneda

Display Applications of PLZT Ceramics

J. R. Maldonado, D. B. Fraser, and A. H. Meitzler

Striped Color Encoded Single Tube Color Television Camera Systems

J. J. Brandinger, G. L. Fredendall, and D. H. Pritchard

List of Contributors

Numbers in parentheses indicate the pages on which the authors' contributions begin.

J. J. BRANDINGER (169), RCA Laboratories, Princeton, New Jersey

D. B. FRASER (65), Bell Laboratories, Murray Hill, New Jersey

G. L. FREDENDALL (169), RCA Laboratories, Princeton, New Jersey

J. R. MALDONADO (65), Bell Laboratories, Murray Hill, New Jersey

A. H. MEITZLER* (65), Bell Laboratories, Murray Hill, New Jersey

D. H. PRITCHARD (169), RCA Laboratories, Princeton, New Jersey

TEIICHI TANEDA (1), NHK (Japan Broadcasting Corporation), Technical Research Laboratories, Kinuta, Setagaya, Tokyo, Japan

MANABU YAMAMOTO (1), Central Research Laboratory, Hitachi, Ltd., Kokubunji, Tokyo, Japan

* Present address: Scientific Research Staff, Ford Motor Company, Dearborn, Michigan.

Preface

Each of the topics included in the present volume has received major attention over the past number of years. In view of the extensive literature which has accumulated in each of these areas it is believed useful at present to provide a unified treatment of these subjects to place them in proper perspective. At the same time the chapters provide a critical review and up-to-date survey of the field for those having a serious interest in the subject. It is hoped the chapters will also serve to provide a useful introduction for those not familiar with the field.

The first article by M. Yamamoto and T. Taneda, entitled "Laser Displays," covers a subject that has received attention almost from the inception of the laser itself. Because of the narrow beam and high power density obtainable from the laser, numerous workers have explored its use for generating TV images. Since a scanned laser beam, unlike an electron beam, need not be confined to the evacuated space of a glass envelope, it allows the generation of images of very large size. In addition, because of the spectral purity of the light obtainable from a laser it is possible, using multiple lasers, to obtain color images whose spectral range exceeds that obtainable with conventional cathode-ray tubes. The idea of generating a TV image with a scanning light beam is itself not new and goes back to the 1930's when mechanical scanners and acoustic deflectors were studied in conjunction with the light sources then available. The recent interest in laser displays has thus served to revive interest in a number of these scanning and modulating systems. As discussed in the present chapter, extremely high quality large-screen color images can be generated with laser systems. However, an important obstacle limiting wider interest in this field at present is the low efficiency of the laser in converting electrical power to light.

The second chapter, entitled "Display Applications of PLZT Ceramics," by J. R. Maldonado, D. B. Fraser, and A. H. Meitzler is based upon the use of a new type of electrooptic material in the form of a ceramic plate. Despite their microcrystalline structure such materials are capable of efficient light modulation, a property usually associated only with single-crystal materials. Since in many cases the ceramic material has ferroelectric properties it can also store electrical information, acting as a light valve whose transmission can be set at an arbitrary level. By using a photoconductive layer to control the ceramic plate, optical input images can thus be recorded and then viewed for indefinite periods.

The third chapter, entitled "Striped Color Encoded Single Tube Color Television Camera Systems," by J. J. Brandinger, G. L. Fredendall, and D. H. Pritchard, covers a topic that has been of interest to workers in the television field for many years. Since the earliest days of color TV broadcasting, camera systems have depended on the use of three separate pickup tubes, each responsive to a different portion of the spectrum. Aside from their bulk and cost, such systems depend on optical registration of the individual color images as well as proper tracking of the electron beams in the separate tubes. The value of a single-tube camera capable of generating the individual color signals has thus been recognized for some time. As discussed in this chapter, a considerable degree of success has been achieved in this direction, single-beam cameras having been developed capable of good performance and relatively low cost. Although their performance is not quite as high as that of the multiple-tube camera, the single-tube cameras are likely to find increasing use in industrial TV systems where compactness and cost are important factors.

B. KAZAN

CONTENTS OF VOLUME 1

Advances in
Image Pickup and Display

VOLUME 2

Laser Displays

Manabu Yamamoto

CENTRAL RESEARCH LABORATORY
HITACHI, LTD.
KOKUBUNJI, TOKYO, JAPAN

and

Teiichi Taneda

NHK (JAPAN BROADCASTING CORPORATION)
TECHNICAL RESEARCH LABORATORIES
KINUTA, SETAGAYA, TOKYO, JAPAN

1

I. Introduction

The introduction of the laser as a source of coherent, visible light has stimulated the development of a new display technology—the laser display. The concept of the laser display is based on the generation of a visual image by means of a scanned, intensity-modulated laser beam. In its simplest form, the laser display system consists of three major parts: the laser light source, the light modulator, and the beam deflector. The light modulator imparts time-varying video signals to the laser beam, with the beam deflector effecting the two-dimensional spatial positioning of the beam.

The laser as a coherent light source has unique features which enable generation of high-quality pictures. The fine spot of the focused beam permits generation of high resolution images. The spectral purity of the light provides a color gamut that is not attainable by conventional light sources. In addition to these benefits, the coherent property of the laser allows for development of new techniques and components for control of the intensity and spatial positioning of a light beam. Present laser display technology relies upon these features of the coherent light source.

Various approaches to a laser display system may be considered (Crocetti *et al.*, 1968). One approach is the direct-view laser display in which the visible laser beam supplies its own light to the screen on which the image is generated. In another approach, the laser beam is used only to control the local optical parameter (reflectivity or transmissivity) of some material. This material acts as a light valve, with a separate conventional light source providing the light to the screen.

The first approach mentioned will be of major concern in this paper, although the discussions presented here are applicable to other concepts. In general, the description is concentrated on display systems with a raster-scanning format such as the television display. With this format, a continuous-wave (CW) laser beam scans the entire image area in a sequential-access format. With the other format, the beam is steered from one direction to another to provide random access. This method is adopted, for example, in laser pattern generation systems (see Section IX, B).

In the following description we will adhere to common television terminology. For example, the term "vertical" means the direction perpendicular to the scanning lines, regardless of the orientation of the scanning lines themselves. The vertical dimension of the picture is called the picture height. The resolution of an image is expressed in the number of resolvable spots per scanning line. The horizontal resolution of N resolvable spots per scanning line means that the system is capable of generating $2N$ alternate black and white dots, or $2N$ picture elements, along the scanning line. The active scanning line is defined as that portion of the scanning line which contains picture information. This is equal to the total scanning line minus the horizontal retrace. It is a common practice in television engineering to express image resolution in TV lines per picture height. For a picture with the aspect ratio of $\frac{3}{4}$, horizontal resolution of N' TV lines per picture height is equivalent to $(\frac{4}{3})N'$ TV lines per scanning line.

II. Historical Summary

The discovery of the red light from a helium–neon laser (White and Rigden, 1962) stimulated interest in applications as well as the search for additional visible transitions. The first display system that used a He–Ne laser was operated in late 1964 by Texas Instruments, Inc. and produced a red and black image from a commercial television broadcast (Baker and Rugari, 1965, 1966; Baker, 1968, 1970). Laser intensity modulation was performed by a potassium dihydrogen phosphate (KDP) electrooptic light modulator. High-frequency horizontal deflection was generated by an acoustically resonant, nutating-mirror scanner. Vertical scanning was accomplished by a galvanometer-driven moving mirror.

The discovery of the rare gas ion laser (Bridges, 1964; Convert et al., 1964; Bennett et al., 1964; Gordon et al., 1964) should be noted as particularly significant, since it provided the best source of high-power visible laser transitions that could be obtained on a continuous basis.

A color television display using tricolor lasers was reported by General Telephone and Electronics Laboratories, Inc. (Stone, 1967). An argon ion laser was used to provide blue and green primaries and a He–Ne laser furnished the red primary. The horizontal and vertical scanning was performed by multiple-element, piezoelectrically driven, moving-mirror scanners. In the improved version of this system, the horizontal scanning was accomplished by a high-speed rotating polygonal mirror, and the He–Ne laser was replaced by a krypton ion laser (Stone et al., 1969; see also Fowler, 1968). A similar device was developed by Alsabrook and Baker (Baker, 1968; Alsabrook, 1966; Baker and Alsabrook, 1967).

Work at the Zenith Radio Corporation opened the way toward a system using acoustic devices in both the modulation and deflection of the laser beam (Korpel et al., 1966). This approach was, as they stated, reminiscent of the old Scophony television display (Okolicsanyi, 1937; Lee, 1938; Robinson, 1939; Sieger, 1939; Wikkenhauer, 1939; Lee, 1939). The modern Zenith approach stimulated interest in the application of the acoustooptic effect to the laser display. The Zenith acoustooptic system was extended to full color display by adding deflection angle multipliers, which were needed to correct the green and blue beam scanning angles for registration with the red beam scanning angle (Watson and Korpel, 1969; Hrbek et al., 1970, 1971).

In the earlier laser display systems the picture area was limited to about 1 m^2 by the optical output capability of the ion laser. It was known that operation of the rare gas ion laser in high discharge current regions suffered from discharge instability (Marantz et al., 1969) as well as rapid degradation of the optical components (DeMars et al., 1968). High-power ion laser technology was developed at the Central Research Laboratory, Hitachi, Ltd., where krypton and argon ion lasers were built which generated optical power of 7 W (CW) on each of the three primary colors (Yamamoto et al., 1971). Using these lasers, Hitachi, Ltd. demonstrated a large screen laser color-TV projector at the 1970 International Exposition held in Japan (Yamada et al., 1970). This system produced television pictures of 10-m^2 area from standard broadcasts. The white beam power was 12 W under normal operating conditions.

At the same time, the NHK (Japan Broadcasting Corporation) Technical Research Laboratories had been trying to develop new display techniques to generate a high-quality television image. A 1125-scanning-line raster was generated by using a high-speed rotating-mirror scanner and a raster irregularity compensator. A wideband video amplifier and an automatic bias controller were incorporated in the electrooptic (EO) modulating system. As a result, horizontal resolution as high as 1500 TV lines per picture height was obtained (Taneda et al., 1973a).

Since 1971, NHK and Hitachi have been collaborating in the development of a large screen, 1125-scanning-line laser color-TV display (Taneda et al., 1973b; Yamamoto, 1975a). This system will be described in Section VIII.

In recent years, work has been continuing on the use of scanned beam techniques for such applications as image scanning and recording, pattern generation, etc. These applications will be surveyed in Section IX.

III. Description of System

In the laser display a scanned image is produced in the same manner as in a cathode-ray tube (CRT) (Baker, 1968). As shown in Fig. 1, a thin beam of light emerging from a continuously operating visible-wavelength laser is intensity-modulated by a light modulator in response to the video signal applied to it. The modulated light beam is then scanned to form a two-dimensional raster on a viewing screen. Compared to the cathode-ray tube which has a glass faceplate coated with phosphor to convert the energy of the electron beam to visible light, the viewing screen for a laser display can be any screen of the kind used for viewing projected slides or motion pictures. Also, in comparison with the CRT, the laser display does not require a vacuum envelope or a special phosphorescent screen. Thus, the limit on screen size imposed by the glass envelope does not apply to laser displays. The lack of screen persistence in a laser display has little effect on direct observation (Easton et al., 1967).

Figure 2 illustrates a three-color laser display system. The lasers generate the three primary color beams, which are intensity-modulated by respective light modulators. The red (R), green (G), and blue (B) video signals are derived from a television receiver or camera. The three modulated beams must now be scanned and projected onto the screen. This can be accomplished by recombining the three beams on a common axis, and then scanning by means of a horizontal and vertical scanner. In this system the angles of beam deflection imparted to the three beams must be precisely the same amount for any deflection angle. This is true for nondispersive deflectors such as moving mirrors. With these achromatic deflectors, essentially perfect color registration is insured.

On the other hand, deflectors based on variable diffraction or refraction of light—i.e., acoustooptic and electrooptic devices—produce wavelength-

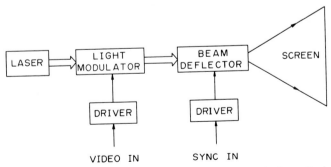

Fig. 1. Schematic diagram showing major components of a laser display.

Manabu Yamamoto and Teiichi Taneda

VIDEO SIGNAL

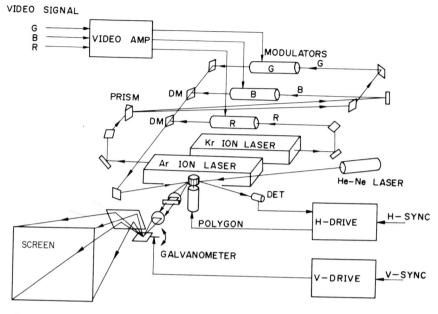

Fig. 2. Schematic diagram of a three-color laser display with a krypton ion laser providing the red (R) primary color and an argon ion laser providing the blue (B) and green (G) primary colors. DM: Dichroic mirorrs. DET: Photodetector.

dependent scanning angles. With these systems, beams of different colors are deflected by separate scanners and are separately directed to a remote screen on which the registration of the three-color images is accomplished. Obviously, a nondispersive deflector is preferable in multicolor display because of the simplicity of the optical system and ease in obtaining registration.

IV. Preliminary Considerations on Resolution

One of the most important parameters that define the picture quality is resolution capability. In the scanned laser beam system the resolution is primarily determined by physical optics. This is because the spot of a laser beam projected on an image plane has a finite spread whose diameter is determined by the diffraction of light.

We begin with the assumption of an aberration-free focusing system and an ideal Gaussian beam with an intensity profile proportional to $\exp[-2(R/R_0)^2]$; where R is the distance from the beam axis and R_0 is the beam radius at which the intensity falls off to e^{-2} of its value on the

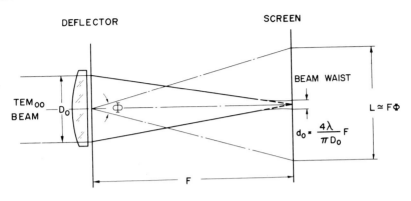

Fig. 3. Deflection of a focused laser beam.

axis. The beam diameter D_0 is defined by $D_0 = 2R_0$. As shown in Fig. 3, this beam can be focused to form a beam waist (or focus), thereby preserving the Gaussian intensity profile. The beam diameter $D(Z)$ at a distance Z from the beam waist is (Yariv, 1967)

$$D(Z) = \{d_0^2 + [(4\lambda/\pi d_0)Z]^2\}^{1/2} \tag{1}$$

where d_0 is the waist diameter measured at the e^{-2} intensity points, and λ is the optical wavelength. The depth of focus is given by the confocal parameter b defined by $b = 2(\pi/4)d_0^2/\lambda$. At distances $Z = \pm b/2$, the beam diameter has increased by a factor of $\sqrt{2}$ over its value at the beam waist. For larger values of Z the second term under the radicant in Eq. (1) dominates. Therefore, the beam becomes conical in shape with an apex angle $\theta = 4\lambda/\pi d_0$. This is the minimum divergence angle to which a Gaussian beam of diameter d_0 can be collimated.

Now suppose that a Gaussian beam of diameter D_0 is deflected through an angle Φ. If the beam is collimated at the deflecting point, the scanned beam would generate a far-field pattern whose resolution is given by

$$N_R = \Phi/\theta = \pi\Phi D_0/4\lambda, \quad \text{with} \quad \theta = 4\lambda/\pi D_0 \tag{2}$$

where θ is the diffraction angle. Equation (2) shows that the resolution attainable by a scanned beam system is proportional to ΦD_0, the product of the deflection angle and the beam diameter.

When the beam is focused by an optical system (as shown in Fig. 3) the resolution obtained on the focal plane is given by the ratio of the scanning-line length, $L \simeq F\Phi$ to the beam spot diameter $d_0 = 4\lambda F/\pi D_0$. Thus, the resolution obtained on the focal plane is identical to that defined by Eq. (2). However, this argument is rigorously correct only for an optical system for which $F \gg b$; i.e., $D_0 \gg d_0$. The general case of a diffraction-limited beam converged through a deflector was discussed by Fowler (1974). He showed that the maximum resolution is obtained on a projection screen located by an additional distance Z_2 behind the beam waist, which is given by

$$Z_2 = (\pi d_0{}^2/4\lambda)^2/Z_1 \qquad (3)$$

where Z_1 is the distance between the focal point and the deflector.* The maximum resolution obtained at this location can be shown to be the same as the value given by Eq. (2). For Z_1 much larger than b, then Z_2/Z_1 approaches zero. As a numerical example, if we let $\lambda = 514.5$ nm, $D = 1$ mm, and $\Phi = 0.5$ rad, then $N_R = 750$ resolvable spots per scanning line. Thus, if the aperture width of the beam deflector is expanded to 10 mm, N_R will equal 7500 resolvable spots per scanning line.

The above discussions are based on the assumption of an ideal Gaussian profile for the beam-intensity distribution. In an actual display system, the light beam is truncated by the finite aperture width of the optical component. In this case, the spatial intensity variations that result when a beam of given intensity profile is scanned in a linear fashion, and simultaneously modulated by a periodic signal, must be calculated (Zworykin and Morton, 1954; Gorog *et al.*, 1972a; Cohen and Gorog, 1973). In the following calculations we assume that the beam is scanned linearly in the x direction on the image plane and that there is no y variation. The beam is assumed to be symmetrical around its central axis. The irradiance distribution of the beam on the image plane centered at $x = 0$ is described by the beam spread function $S(x)$, which is normalized to unity: $\int_{-\infty}^{+\infty} S(x)dx = 1$. If we assume that the beam is scanned at a constant linear velocity v, then the instantaneous irradiance distribution becomes $S(x - vt)$, as shown in Fig. 4. If the beam intensity is modulated by $M(t)$, the instantaneous exposure is $M(t)S(x - vt)$, and the total exposure is

$$E(x) = \int_{-\infty}^{+\infty} M(t)S(x - vt)dt \qquad (4)$$

In direct-view display systems, integration is accomplished by the persistence of vision of the human eye. Assume that the beam intensity is modulated sinusoidally at the angular frequency ω; i.e., $M(t) = 1 + A \cos(\omega t)$ for $A < 1$. Then from Eq. (4) we have

$$E(x) = \int_{-\infty}^{+\infty} S(x - vt)[1 + A \cos(\omega t)]dt \qquad (5)$$

---

* This result is obtained by calculating Z that maximizes $N_R(Z) = (Z_1 + Z)\Phi/D(Z)$.

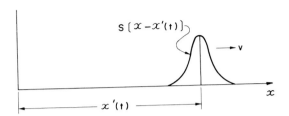

Fig. 4. Intensity distribution of a scanned spot. $S(x-x'(t))$: intensity profile of a spot centered at $x'(t)$. For linear scanning at a constant velocity v, $x'(t) = vt$.

By making $x - vt = \tilde{x}$ and noting that $S(-x) = S(x)$, we obtain

$$E(x) = v^{-1} \int_{-\infty}^{+\infty} S(\tilde{x})[1 + A \cos (kx - k\tilde{x})]d\tilde{x}$$

$$= v^{-1}[1 + AR(k) \cos (kx)] \tag{6}$$

where the frequency response function $R(k)$ is given by

$$R(k) = \int_{-\infty}^{+\infty} d\tilde{x} S(\tilde{x}) \cos (k\tilde{x}) \tag{7}$$

and $k = \omega/v$ is the spatial angular frequency. The last formula shows that the frequency response of a modulator-scanner system is given by the Fourier transform of the spread function $S(x)$.

When the limiting aperture of the optical system W is much larger than the incident beam diameter $D_0 (W \gg D_0)$, the spread function $S(x)$ is closely approximated by the Gaussian profile treated above:

$$S(x) = (2/\pi)^{1/2} r_0^{-1} \exp (-2x^2/r_0^2) \tag{8}$$

where $r_0 = d_0/2$ is the radius of the scanning spot at the e^{-2} intensity point. In this case the frequency response is also Gaussian:

$$R(k) = \exp [- (kr_0/2\sqrt{2})^2] \tag{9}$$

In the opposite case, i.e., when $W \ll D_0$, we have a far-field diffraction pattern produced by a uniformly illuminated slit of width W:

$$S(x) = \pi^{-1} \sin^2 (\pi Wx/\lambda F)/(\pi Wx/\lambda F)^2 \tag{10}$$

In this case we obtain

$$F(k) = 1 - k/k_{max} \qquad k \leq k_{max}$$

$$= 0 \qquad k > k_{max} \tag{11}$$

where $k_{max} = 2\pi W/\lambda F$. Since $k_{max}/2\pi$ is the maximum number of resolvable spots per unit of scanning length, the resolution can be expressed as

$$N_R = (k_{max}/2\pi)\Phi F = \Phi W/\lambda \tag{12}$$

Results of numerical calculations for the intermediate range, $D_0 \approx W$, are shown in Fig. 5 with the truncation ratio $\rho = W/D_0$ as a parameter (Korpel *et al.*, 1971; Randolph and Morrison, 1971; Beiser, 1974). Since the illumination function is peaked at the center, the spread function will broaden. Therefore, the high-frequency response with Gaussian illumination will always be poorer than with uniform illumination. Figure 5 shows that the limiting resolution is always given by Eq. (11), irrespective of the truncation ratio W/D_0. This result can be understood from the generalized expression for $R(k)$ (Gorog *et al.*, 1972a; Cohen and Gorog, 1973):

$$R(k) = \int_{-\infty}^{+\infty} d\tilde{x} e^{-ik\tilde{x}} S(\tilde{x})$$

$$= \int_{-\infty}^{+\infty} dx_0 P(x_0) P^*(x_0 - \lambda F k/2\pi) \tag{13}$$

where $P(x_0)$ is the amplitude distribution of the incident light across the

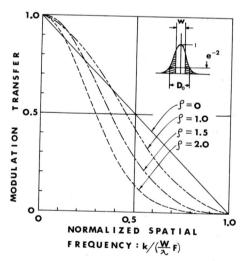

Fig. 5. Modulation transfer functions versus relative spatial frequency for a one-dimensional Gaussian beam of width D_0 (at e^{-2} intensity points) truncated by a slit of width W. The truncation ratio is defined by $\rho = W/D_0$ (Korpel *et al.*, 1971; Beiser, 1974.)

aperture. Since $P(x_0) = 0$ for $| x_0 | > W/2$, we see $R(k) = 0$ for $k > 2\pi W/(\lambda F)$.

V. Coherent Light Sources

A. Various Types of Lasers

Lasers used as the light source for scanned beam display must be capable of stable, continuous-wave operation in the visible spectrum region. The output beam of the uniphase (TEM$_{00}$) mode is preferred. The oscillation wavelength must have a good spectral match with the primary colors of the standard television system, or with the photosensitization curve of the recording medium.

The most useful laser for large screen display has been the rare gas ion laser. The ion laser medium is typically low pressure ($\sim$1 torr) argon or krypton gas in a small bore (1–5 mm) discharge tube. When a current of 10–50 A is passed through the capillary, rare gas atoms are excited to high energy states (Kitaeva *et al.*, 1974), and intense laser emissions are observed which belong to the np–ns transitions of the singly ionized rare gas atom. The krypton ion laser provides the red primary, KrII 647.1 nm

Fig. 6. High-power rare gas ion laser developed as a light source for the 1125-scanning-line laser color-TV display system. The optical cavity length is 110 cm.

Manabu Yamamoto and Teiichi Taneda

$(4p^45s\ ^2P_{3/2}\ -\ 4p^45p\ ^4P^\circ_{5/2})$, while the argon ion laser provides the green primary, ArII 514.5 nm $(3p^44s\ ^2P_{3/2}\ -\ 3p^44p\ ^4D^\circ_{5/2})$, and the blue primary, ArII 476.5 nm $(3p^44s\ ^2P_{1/2}\ -\ 3p^44p\ ^2P^\circ_{3/2})$.

Ion laser technology has progressed through several stages with corresponding increase in power output and tube life (Geusic *et al.*, 1970; Kitaeva *et al.*, 1970). Commercially available models generate multiwatt optical power at the blue or green line of argon, or at the red line of krypton. Figure 6 shows an experimental model built for use as the light source of a recently developed, 1125-scanning-line display system (see Section VIII). The active region of this laser is a discharge tube made of a beryllium oxide capillary with a 2- to 3-mm bore and 400 mm in length. A longitudinal magnetic field of 500–600 gauss is applied to the discharge to improve optical efficiency. The gas-filling pressure is 1–2 torr. Figures 7a and 7b show the optical output characteristics of an argon and a krypton ion

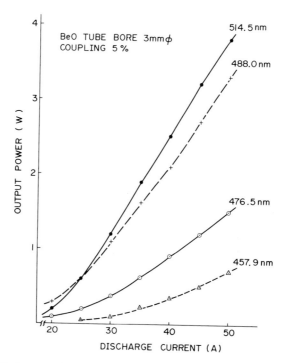

Fig. 7a. Optical output characteristics of an argon ion laser used in the 1125-scanning-line laser color-TV display system. Power was measured in a single-line oscillation mode, which was obtained using a Littrow mirror as an optical cavity component. Coupling percent indicated corresponds to the transmission coefficient of the output mirror.

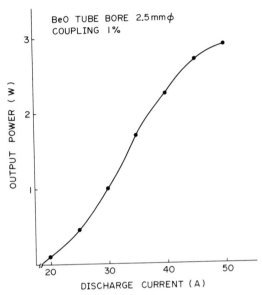

Fig. 7b. Optical output characteristics of a krypton ion laser used in the 1125-scanning-line laser color-TV display system.

laser, respectively. The laser is normally operated at an electric power input of 12 kW, for which the optical output powers are 2 W (KrII 647.1 nm), 3 W (ArII 514.5 nm), 3 W (ArII 488.0 nm), and 1 W (ArII 476.5 nm). The output beam is in low order multimode, but TEM_{00} mode operation can be obtained at a sacrifice of output power.

The practical limitation on the optical power available from this laser is set by a discharge instability, which begins suddenly when the discharge current exceeds a critical point. An empirical formula for the condition of stable operation at an externally applied magnetic field of 500–600 gauss is

$$p > (1.7 \pm 0.4) \times 10^{-3} j, \tag{14}$$

where p is gas-filling pressure in torr and j is current density in A/cm^2.

When operated with an argon–krypton gas mixture, the rare gas ion laser generates a white luminous flux composed of three color components (Hrbek et al., 1970; Ahmed and Keeffe, 1974). Mixed-gas rare gas ion lasers are commercially available.

The He–Ne laser has red output at 632.8 nm, and the He–Cd laser has blue output at 441.6 nm (Silfvast, 1968). Two-color output can be obtained from a He–Ne–Cd laser (Ahmed and Campillo, 1969), while the He–Se laser has many lines in the visible region (Silfvast, 1970). These lasers have low output power characteristics (1–100 mW), and are there-fore suited for such applications as small scale display systems, image scanning and recording, etc.

The Nd:YAG laser is an optically pumped solid-state device which can be operated on a continuous basis (Geusic *et al.*, 1970). The infrared out-put at 1.06 nm can be converted to its second harmonic wave at 0.53 nm by using a nonlinear optical crystal such as barium sodium niobate ($Ba_2NaNb_5O_{15}$), as shown in Fig. 8. A second harmonic output of 0.35 to 0.5 W has been reported for a pumping lamp input of 800 W (Culshaw *et al.*, 1974; see also Murray *et al.*, 1974). The efficiency would be im-proved by using a high-pressure potassium vapor lamp as the pumping light source, which has an emission spectrum in better agreement with the absorption bands of Nd:YAG (Liberman *et al.*, 1969).

Table I lists some of the important laser lines suited for display ap-plications.

B. COLOR AND LUMINANCE

The laser has inherently narrow spectral lines. When no special mode selection means are incorporated, the linewidth of a gas laser is determined by Doppler broadening, which is $1.5 - 7$ GHz, or one part in 10^5 of the optical frequency. The width of the frequency spectrum of a typical multi-axial mode Nd:YAG laser is 15 to 30 GHz. Since the perception of color differences requires frequency change of a few percent, the laser offers

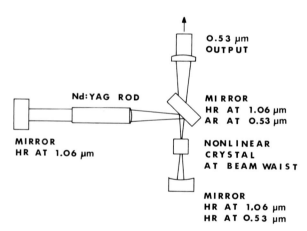

Fig. 8. Intracavity frequency converted, unidirectional output Nd:YAG laser. Second harmonic generation (SHG) can be obtained using a nonlinear crystal such as barium sodium niobate (BNN), cesium dihydrogen arsenate (CDA), lithium iodate, etc. HR: high reflection, AR: anti-reflection.

TABLE I

PRIMARY COLORS RELATED TO LASER DISPLAY

Color	Wavelength (nm)	Chromaticity coordinates		Luminous efficiency of radiated power (lm/W)
		x	y	
Red	Kr 647.1	0.72	0.28	85
	He–Ne 632.8	0.71	0.29	162
	NTSC R	0.67	0.33	342
Green	Nd:YAG 532	0.17	0.80	600
	Ar 514.5	0.04	0.81	406
	NTSC G	0.21	0.71	622
Blue	Ar 488.0	0.05	0.25	131
	Ar 476.5	0.10	0.10	82
	He–Cd 441.6	0.16	0.01	17
	NTSC B	0.14	0.08	68

pure, fully saturated colors. Figure 9 shows the chromaticity coordinates of laser lines. The solid triangle shows the color gamut of a standard television broadcast system. It is evident that by an appropriate choice of the laser primaries, a wider range of colors can be reproduced than can be achieved with other color display systems.

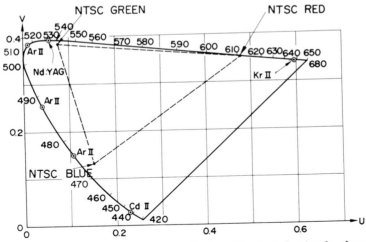

Fig. 9. Chromaticity diagram for laser display. The dashed triangle shows the color gamut of the NTSC primary colors.

According to the theory of vector addition of colors, a white luminous flux **C** can be matched by three basic color vectors, **R** (red), **G** (green), and **B** (blue):

$$\mathbf{C} = r\mathbf{R} + g\mathbf{G} + b\mathbf{B} \tag{15}$$

where the basic vectors are normalized to unit luminance and the coefficients r, g, and b represent luminance contributions of the three primaries that make up **C**. We take the CIE standard illuminant "C" as the white flux **C** and put $|\mathbf{C}| = 100$ lm. Now, we can calculate the contributions of the three primary colors. Table II shows components of a 100-lm white flux **C** for various combinations of the three primary colors.

Luminance of a screen when it is scanned by a laser beam can be calculated by assuming that instantaneous luminance is averaged out by the persistence of vision of the human eye. Denoting the screen luminance by B and the luminous flux by Y, we obtain

$$B = (1/\pi S)YG(1 - \xi_{\mathrm{H}})(1 - \xi_{\mathrm{V}})\epsilon \tag{16}$$

where S is the active picture area—i.e., excluding the blanking—and G is the screen gain representing directional property of reflectance of a surface. (The screen is G times brighter than the Lambertian surface when observed in the direction normal to the surface.) The horizontal and vertical blanking factors are represented by ξ_{H} and ξ_{V}, respectively. The transmission coefficient of the optical system is denoted by ϵ.

To give a numerical example, we put $S = 3\mathrm{m}^2$, $G = 4$, $\xi_{\mathrm{H}} = 0.16$, $\xi_{\mathrm{V}} = 0.08$, and $\epsilon = 0.3$. With the optical power of 2 W at 647.1 nm, 0.75 W at 514.5 nm, and 0.87 W at 476.5 nm, we have a white flux of $Y = 550$ lm. By substituting these values into Eq. (16) we get $B = 54$ nit $= 16$ fL.

C. Comparison of Coherent and Incoherent Sources

All the known lasers that are capable of CW operation in the visible region have a very low efficiency in converting electrical energy to light. The luminous efficiency of a rare gas ion laser is typically of the order of 10^{-2} lm/W. On the other hand, we know that conventional lighting sources have efficiencies of 20 to 100 lm/W. The question naturally arises as to the reason these incoherent sources cannot be used as the light source for scanned beam display. This question will be answered by showing numerically that the extremely high level of luminance that is required of the light source can be attained only by the use of a coherent light source.

Let us suppose that a light flux radiated from a finite area of a radiating surface was collimated with the appropriate optics. The collimated beam would have an unavoidable divergence, the angle of which depended on the area of the radiating source. The source area would thus have to be

TABLE II

COMPONENTS OF WHITE LUMINOUS FLUX (CIE "C") MADE UP OF VARIOUS COMBINATIONS OF THREE COLORS

Green	Blue	Red: Kr 647.1 nm			He-Ne 632.8 nm			NTSC R		
		Wavelength (nm)	Flux (lumen)	Power (W)	Wavelength (nm)	Flux (lumen)	Power (W)	Wavelength (nm)	Flux (lumen)	Power (W)
Ar 514.5 nm	Ar 488.0 nm	647.1	33	0.39	632.8	36	0.22	**R**	44	0.13
		514.5	25	0.06	514.5	22	0.06	514.5	14	0.03
		488.0	42	0.32	488.0	42	0.32	488.0	43	0.32
		Total	100	0.77		100	0.59		100	0.48
	Ar 476.5 nm	647.1	31	0.36	632.8	33	0.20	**R**	40	0.12
		514.5	56	0.14	514.5	54	0.13	514.5	46	0.11
		476.5	13	0.16	476.5	13	0.16	476.5	13	0.16
		Total	100	0.66		100	0.49		100	0.39
	He-Cd 441.6 nm	647.1	28	0.33	632.8	30	0.19	**R**	37	0.11
		514.5	70	0.17	514.5	68	0.17	514.5	62	0.15
		441.6	1.5	0.09	441.6	1.5	0.09	441.6	1.5	0.09
		Total	100	0.59		100	0.44		100	0.35
	NTSC **B**	647.1	29	0.34	632.8	31	0.19	**R**	38	0.11
		514.5	60	0.15	514.5	58	0.14	514.5	51	0.13
		B	11	0.16	**B**	11	0.16	**B**	11	0.16
		Total	100	0.63		100	0.48		100	0.38

(Continued)

TABLE II—*Continued*

Green	Blue	Red: Kr 647.1 nm Wavelength (nm)	Flux (lumen)	Power (W)	He–Ne 632.8 nm Wavelength (nm)	Flux (lumen)	Power (W)	NTSC R Wavelength (nm)	Flux (lumen)	Power (W)
Nd:YAG (2nd Harmonic) 532 nm	Ar 488.0 nm	647.1	32	0.37	632.8	34	0.21	**R**	42	0.12
		532.	25	0.04	532.	23	0.04	532.	15	0.02
		488.0	43	0.33	488.0	43	0.33	488.0	43	0.33
		Total	100	0.74	—	100	0.57	—	100	0.47
	Ar 476.5 nm	647.1	27	0.31	632.8	29	0.18	**R**	36	0.10
		532.	59	0.10	532.	57	0.10	532.	50	0.08
		476.5	14	0.17	476.5	14	0.17	476.5	14	0.17
		Total	100	0.58	—	100	0.44	—	100	0.36
	He–Cd 441.6 nm	647.1	22	0.26	632.8	24	0.15	**R**	30	0.09
		532.	76	0.13	532.	74	0.12	532.	69	0.11
		441.6	1.6	0.10	441.6	1.6	0.10	441.6	1.6	0.10
		Total	100	0.49	—	100	0.37	—	100	0.30
	NTSC B	647.1	24	0.28	632.8	26	0.16	**R**	32	0.09
		532.	64	0.11	532.	62	0.10	532.	56	0.09
		B	12	0.17	**B**	12	0.17	**B**	12	0.18
		Total	100	0.55	—	100	0.42	—	100	0.34

NTSC G

Source	Primary	647.1 %	647.1	632.8 %	632.8	R %	R
Ar 488.0 nm	647.1 / 632.8 / R	31	0.36	33	0.20	42	0.12
	G	27	0.04	24	0.04	16	0.03
	488.0	42	0.32	42	0.32	43	0.33
	Total	100	0.73	100	0.56	100	0.47
Ar 476.5 nm	647.1 / 632.8 / R	25	0.29	27	0.16	33	0.10
	G	62	0.10	60	0.10	53	0.09
	476.5	14	0.17	14	0.17	14	0.17
	Total	100	0.56	100	0.42	100	0.35
He–Cd 441.6 nm	647.1 / 632.8 / R	20	0.24	22	0.13	27	0.08
	G	78	0.13	77	0.12	71	0.11
	441.6	1.6	0.09	1.6	0.09	1.6	0.09
	Total	100	0.45	100	0.35	100	0.28
NTSC B	647.1 / 632.8 / R	22	0.26	24	0.15	30	0.09
	G	66	0.11	65	0.10	59	0.09
	B	11	0.17	11	0.17	11	0.17
	Total	100	0.52	100	0.40	100	0.33

small enough to satisfy the resolution requirement given by Eq. (2). In other words, the radiator would have to be of sufficiently high luminance capable of providing the required amount of light from the finite area.

If a beam of luminous flux Y is radiated into a solid angle Ω from a source of area S, the luminance is defined as

$$B = Y/S\Omega \qquad (17)$$

We put $S = W^2$, where W is the aperture width of the optical system. When this beam is scanned through an angle Φ, the limiting resolution is given by $N_R = \Phi/\theta$, where θ is the full angle of unavoidable beam divergence. Using the relation $\Omega = (\pi/4)\theta^2$, we have

$$B = \frac{4}{\pi}\frac{YN_R^2}{S\Phi^2}. \qquad (18)$$

Putting $Y = 500$ lm, $W = 1$ cm, $\Phi = 0.5$ rad, and $N_R = 1000$, we obtain $B = 2.5 \times 10^9$ cd/cm^2 = 7.4×10^{12} fL. We note here that, according to the principle of geometrical optics, light flux can only be focused from a given source to another location that is less bright than the source itself (Born and Wolfe, 1970, Section 4.8; Chang, 1968). Therefore, the above figure is the minimum requirement on the luminance of the light source. We know that incoherent sources such as the xenon short arc lamp have luminance of the order of 10^5 cd/cm^2 and a specially designed high-intensity lamp with luminance of 10^6 cd/cm^2 has been reported (Thouret *et al.*, 1964). It is evident, however, that these lamps are far from capable of satisfying the luminance requirement of scanned beam display. Furthermore, luminance of the incoherent source would be further reduced if the emission was restricted to narrow spectral band monochromatic light.

Now, let us calculate the luminance of a coherent source. We assume a Gaussian beam of diameter D_0 (e^{-2} intensity points), and put $S = (\pi/4)D_0^2$ and $\theta = 4\lambda/\pi D_0$. From Eq. (17) we obtain

$$B = Y/\lambda^2 \qquad (19)$$

If $Y = 500$ lm and $\lambda = 514.5$ nm, then $B = 1.9 \times 10^{11}$ cd/cm^2. Obviously, the luminance requirement is fully satisfied by the coherent source.

As another example, let us consider a high resolution laser recording system, in which an image is recorded on 100-cm^2 photosensitive film in 10 s with a 441.6-nm He–Cd laser beam. The film used is supposed to be of the silver halide type with a sensitivity of 1 erg/cm^2. This would make $Y = 1$ μW in Eq. (18). We suppose that $\Phi = 1$ rad and require that $N_R = 10^4$ resolvable spots per picture width. Under these conditions we obtain $B = 5.1 \times 10^2$ W·cm^{-2}·sr^{-1} as the minimum radiance requirement. This is obviously beyond the capability of incoherent sources, since conventional lamps of the highest intensity can generate radiance of the order of 10 $W \cdot cm^{-2} \cdot sr^{-1}$ per monochromated spectral width of 1 nm (Cann, 1969).

VI. Light Modulation

A. General Remarks

The function of the light modulator is to impose time-varying signals on a light beam resulting in an intensity variation of the beam as a function of time. When the motion imparted to the beam by the deflector is synchronous with the video signal, the time-varying signal is converted to a spatial pattern on the image plane. The modulator characteristics of primary importance in the display application are bandwidth, transmission coefficient, gray-scale reproduction, contrast ratio, and driving power requirements.

The bandwidth required in a modulation system is a function of the image quality and accessing format of the display. In a sequential accessing display in which the beam is scanned linearly with a constant velocity, the horizontal resolution N_R and bandwidth ν are related as follows:

$$N_R = \nu H (1 - \xi_H) \quad \text{(resolvable spots per active scanning line)} \quad (20)$$

where H is the horizontal scanning time. By denoting the number of scanning lines per frame as n_s and the number of frames per second as n_f, we get $H = (n_s \cdot n_f)^{-1}$. Substituting this into Eq. (20) we obtain

$$\nu = \frac{N_R n_s n_f}{(1 - \xi_H)} \quad (21)$$

A typical example might have $N_R = 500$, $n_s = 1000$, $n_f = 30$, and $\xi_H = 0.16$; in this case, ν would equal 18 MHz.

The transmission coefficient of the modulator is important to the maximum utilization of available laser power and to stable operation of the modulator. The power of the incident beam is dissipated through Fresnel reflection at the modulator interfaces, absorption in the modulator crystal, and scattering. Use of materials with low absorption coefficients is of primary importance since even a small percentage of absorption at high laser power levels can cause excessive heating of the modulator elements.

The contrast ratio—or the ratio of the maximum laser power transmitted through the modulator to the minimum value which can be achieved—is very important in the generation of high-quality pictorial images, but is less significant in the presentation of alphanumerics and symbols. A contrast ratio of 30 to 1 is the minimum requirement for pictorial images, and a ratio higher than 50 to 1 is usually recommended.

There are various mechanisms available for intensity modulation of the laser beam. However, only acoustooptical (AO) and electrooptical (EO) techniques have the versatility and the degree of development that satisfy most application requirements. In the region of low modulation frequencies (<10 MHz), the AO modulator is definitely superior to the EO modulator because of the low power required, rugged construction, and compactness (Nowicki, 1974). In the region of medium modulation frequencies (~20 MHz) the EO modulator becomes competitive with the AO modulator. This is because the AO modulator has the disadvantage of a bandwidth limitation imposed by the finite transit time of acoustic waves traversing the light beam. Although the EO modulator is eventually preferred when the frequency response must exceed the bandwidth limitation of the AO device, the image quality obtained with the EO modulator is not always superior to that generated by the AO modulator, since good contrast ratio is very difficult to obtain with the EO modulator.

For both the AO and EO modulators the relationship between output light and input voltages takes a sine-squared form. The relation can be linearized by optical feedback or with a nonlinear amplifier.

B. AO Modulator

The most convenient method of achieving analog modulation of laser light is by means of AO modulators. In an AO modulator, light modulation is achieved through diffraction of light by periodic variation of the refractive index, which is produced by acoustic waves propagating within a transparent medium. An early example of application of acoustic phase grating to light modulation is the television projector developed by Scophony Ltd., London, in the 1930's (see Section II). In this old system, the AO material was a liquid and the light source was a superpressure mercury vapor lamp. For modern AO materials reference is made to the publication by Uchida and Niizeki (1973).

Acoustic diffraction of light is illustrated in Fig. 10. A light beam is incident on a variable-index medium, whose refractive index is described by

$$n(x, t) = n_0 + n_1 \sin (\Omega t - Kx) \qquad (22)$$

In this expression n_0 is the average value of $n(x, t)$, n_1 is the amplitude of the refractive index change and Ω is the angular frequency of sound propagating within the medium in the x direction. The wavenumber K is related to the sound velocity V and wavelength Λ by $K = \Omega/V = 2\pi/\Lambda$. The parameters pertinent to describing AO interaction (Gorog et al., 1972a;

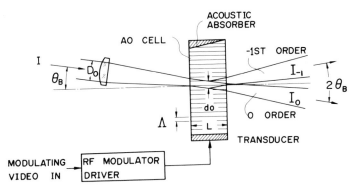

Fig. 10. Acoustooptic (AO) Bragg diffraction. The angle denoted by θ_B is the Bragg angle.

Klein and Cook, 1967; Cohen and Gordon, 1965; Born and Wolf, 1970, Chap. 12; Gordon, 1966) are:

$$\Delta\phi = kn_1 L \tag{23}$$

$$Q = K^2 L/n_0 k \tag{24}$$

$$\alpha = -n_0 k \sin \bar{\theta}/K \tag{25}$$

where the optical wavenumber k is related to the vacuum wavelength λ by $k = 2\pi/\lambda$. The AO interaction length L is the width of the sound wavefront in the plane formed by $\mathbf{k}$ and $\mathbf{K}$, and $\bar{\theta}$ is the incidence angle of light measured within the medium. The first interaction parameter $\Delta\phi$ describes the phase shift, the second parameter Q is the normalized interaction-length parameter, and α is the normalized angle parameter.

In the region $Q \ll \pi$ (the Debye-Sears or Raman-Nath region), the incident light beam is diffracted into many discrete orders of different diffraction angles. If the interaction length is sufficiently large, so that $Q \gg \pi$, nearly all of the incident light is diffracted only into the minus first order when the angle of incidence satisfies the Bragg condition: $|\alpha| = \frac{1}{2}$. That is,

$$n_0 \sin \bar{\theta}_B = \sin \theta_B = K/2k \tag{26}$$

The angles $\bar{\theta}_B$ and θ_B are the Bragg angles measured inside and outside the medium, respectively. In practice, $K \ll k$, and therefore $\sin \bar{\theta}_B \simeq \bar{\theta}_B$, and $\sin \theta_B \simeq \theta_B$.

In Bragg diffraction, the minus first-order beam emerges at an angle $-\theta_B$, as if it is reflected by the acoustic wavefronts. The angular frequency of the emerging light is equal to $\omega \pm \Omega$, where ω is the angular frequency of the incident light, and the plus and minus signs correspond to the direction of sound wave propagation. The intensity of the emerging light I_{-1} is related to the intensity of the incident light I by

$$I_{-1} = I \sin^2 (\Delta\phi/2) \tag{27}$$

This is strictly valid only for the limiting case, $Q = \infty$, which is physically unattainable. However, this expression is approximately correct for many practical cases. For the condition $Q = 2\pi$, the maximum value of I_{-1} was calculated to be 0.975 rather than 1.0 (Hance, 1964; Hance and Parks, 1965). The dependence of I_{-1} on $\Delta\phi$ is found in the literature for the various values of the acoustooptic interaction parameters (Klein and Cook, 1967). The physical meaning of $Q/4\pi$ is the number of acoustic wavefronts traversed by a light ray crossing the acoustic field of width L at the Bragg angle $\bar{\theta}_B$. This is because we obtain from Eqs. (24) and (26) $Q/4\pi = L \sin \bar{\theta}_B/\Lambda$. The peaks of light intensity versus incident angle become sharper with increasing value of $Q/4\pi$, i.e., number of acoustic wavefronts traversed (see Section VII, E).

The phase shift $\Delta\phi$ depends on the acoustic power P_a in the following way:

$$\Delta\phi = \frac{\pi}{\lambda} \left(\frac{2L}{Z} M_2 P_a \right)^{1/2} \tag{28}$$

where Z is the dimension of the sound field perpendicular to both the light and the sound wave propagations; M_2 is the material constant called the acoustooptic figure of merit and is defined by $M_2 = n_0^6 p^2/\rho V^3$, where p is the photoelastic constant and ρ is the density. Some well-known acoustooptic materials are shown in Table III. Suppose that a modulator is made of lead molybdate and has dimensions $L = 10$ mm and $Z = 0.5$ mm. Then the maximum modulation ($\Delta\phi = \pi$) occurs at $P_a = 0.3$ W ($\lambda = 632.8$ nm).

In the AO modulator, the modulated beam can be completely separated from the incident beam because beams of different orders emerge from the modulator in different directions. This fact insures a good contrast ratio for the AO modulator.

The frequency response of an AO modulator can be obtained in a way analogous to that shown in Section IV. For small amplitude modulations the sinusoidal function on the right hand side of Eq. (27) can be expanded to obtain $I_{-1} \sim P_a$. Suppose that an acoustic power P_a is modulated sinusoidally at an angular frequency ω; then the traveling acoustic wave has a power distribution

$$P_a(x, t) = P_0[1 + A \cos (\kappa x - \omega t)],$$

TABLE III

ACOUSTOOPTIC PROPERTIES OF SOME MATERIALS (at λ = 633 nm)[a]

Material	Density (g/cm^3)	Acoustic wave		Optical wave		Figure of merit M_2 $(10^{-18} sec^3/g)$
		Mode and propagation direction	velocity $(mm/\mu s)$	Polarization[b]	Refractive index	
α-HIO$_3$	5.0	Long. [001]	2.44	[100]	1.986	86
PbMoO$_4$	6.95	Long. [001]	3.63	$\parallel$	2.262	36.3
TeO$_2$	6.00	Long. [001]	4.20	$\perp$	2.260	34.5
Fused quartz	2.20	Longitudinal	5.96	$\perp$	1.457	1.56
Te glass	5.87	Longitudinal	3.40	$\parallel$	2.089	23.9
Water	0.997	Longitudinal	1.49	Arbitrary	1.330	126

[a] Values taken from Uchida and Niizeki (1973).

[b] $\parallel$ and $\perp$: parallel and perpendicular to the acoustic wave vector, respectively.

where P_0 is the average power and $\kappa = \omega/V$ is the wavenumber of the modulation envelope. The modulated sound wave travels across a light beam whose intensity profile is given by a normalized spread function $S(x)$. If the beam width is sufficiently large, so that many wavefronts are present within the beam, then the intensity of the diffracted beam at time t is given by

$$I_{-1}(t) \sim \int_{-\infty}^{+\infty} S(x)P_a(x, t)\, dx$$

$$= \int_{-\infty}^{+\infty} S(x)[1 + A \cos (\kappa x - \omega t)]\, dx \qquad (29)$$

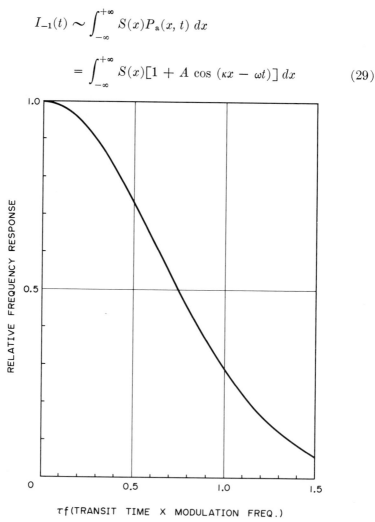

Fig. 11. Relative frequency response of an AO modulator for small-amplitude modulation. The horizontal axis is the product of the transit time $\tau = d_0/V$ and the modulation frequency f.

Comparing this expression with Eq. (6), we find that the frequency response of the AO modulator takes the same form as given by Eq. (7) with k in Eq. (7) replaced by κ. For a Gaussian beam, therefore, Eq. (9) can be used to describe the frequency response of an AO modulator, noting that in this case d_0 is interpreted as the diameter of the light beam interacting with an acoustic wave. Figure 11 shows the relative frequency response of the AO modulator. The half intensity point (-3 dB response) is given by $(\kappa/2\pi)d_0 = f\tau = 0.75$, where $\tau = d_0/V$ is the acoustic wave transit time and f is the modulation frequency. As a numerical example, assume $d_0 = 0.2$ mm and $V = 3.7$ mm/μs (lead molybdate). This results in the -3 dB bandwidth being at 14 MHz. The frequency response can be improved by making τ smaller. However, for very small values of d_0, Eq. (9) is no longer valid. In this region the number of acoustic wavefronts traversing the light beam decreases, and the maximum value of I_{-1} also falls off. When $d_0 \approx \Lambda$, separation between the zero and the minus first-order beams is not assured, since the beam divergence angle then equals angular separation between the different order beams.

Instead of making d_0 very small, there is another approach in which the acoustic transit time is intentionally made long, so that spatial modulation of a light beam is obtained. Along the broadened cross section of the beam, the diffracted light carries a visible replica of the acoustic field, which is spatially modulated in response to a video signal. The image moves at a speed corresponding to the sound propagation. However, the image can be immobilized on the image plane when the whole beam is scanned at the same speed in the opposite direction. Image immobilization by means of a rotating mirror was adopted in the old Scophony television projector. In the modern version of this approach, the Zenith Radio Corporation employed an AO deflector (Korpel *et al.*, 1966).

C. EO Modulator

In the EO modulator, light is modulated through the electric-field induced birefringence of a crystal. For a uniaxial crystal such as potassium dihydrogen phosphate (KDP), the index ellipsoid can be written as (Chen, 1970)

$$n_o^{-2}(x^2 + y^2) + n_e^{-2}z^2 = 1 \qquad (30)$$

where the z axis is along the crystallographic c axis, and n_o and n_e are the ordinary and extraordinary indices of refraction, respectively. The index ellipsoid is shown in Fig. 12 by solid lines.

Application of an electric field to this crystal induces a small change in the shape and orientation of the index ellipsoid. When an electric field E_z

is applied along the crystallographic c axis, the index ellipsoid becomes

$$(n_o + \Delta n)^{-2} x'^2 + (n_o - \Delta n)^{-2} y'^2 + n_e^{-2} z^2 = 1 \qquad (31)$$

where x' and y' are the new coordinate axes in which the deformed ellipsoid is rediagonalized (Fig. 12), and $\Delta n = \frac{1}{2} n_o^3 r_{63} E_z$. The linear electrooptic tensor r_{ij}, which relates the change in the reciprocal dielectric tensor to the applied field, is listed for various classes in the literature (Harvey, 1970).

In a longitudinal modulator, light is propagated along the z axis, and intensity modulation is obtained by the interference of the *normal modes*; that is, light beams polarized along the x' and y' axes. The differential phase delay, or retardation, between these two modes is

$$\Delta\phi = 4\pi \Delta n L_z / \lambda \qquad (32)$$

where L_z is the optical path length and λ is the vacuum wavelength. The maximum modulation occurs at $\Delta\phi = \pi$, for which the applied voltage is given by

$$V_\pi = (E_z L_z)_{\Delta\phi=\pi} = \lambda / 2 n_o^3 r_{63} \qquad (33)$$

which is independent of the size of the crystal. For KDP, $n_o = 1.51$, $r_{63} = -10.5 \times 10^{-12}$ m/V, and $V_\pi = 8.8$ kV ($\lambda = 633$ nm).

Reduction of the half-wave voltage can be achieved in the transverse mode operation, in which the electric field is applied in a direction normal to the light path. This can be realized (as shown in Fig. 13) with an x', y', z rectangular prism of KDP by choosing the light path along the y' axis and the field E_z along the z axis. The incident wave is polarized in the x'-z plane at an angle of $45°$ from the z axis. The retardation between the x' and z components is

$$\Delta\phi = (n_o - n_e + n_o^3 r_{63} E_z / 2)(2\pi L_{y'} / \lambda) \qquad (34)$$

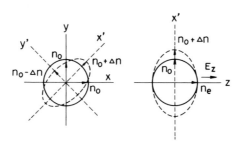

Fig. 12. Index ellipsoid before (solid lines) and after (dashed lines) an electric field E_z is applied.

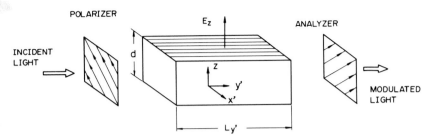

Fig. 13.   Transverse mode electrooptic modulator.

and the half-wave voltage is

$$V_\pi = [\lambda/(n_o^3 r_{63})](d/L_{y'}) \tag{35}$$

where d is the electrode spacing and $L_{y'}$ is the optical path length. The half-wave voltage V_π can be reduced by choosing a smaller geometrical factor $d/L_{y'}$. This can be achieved, however, only at the expense of increased interelectrode capacitance.

Figure 14 shows the output stage of a wideband video amplifier designed as a driver for a transverse EO modulator, whose interelectrode capacitance is 230 pF and $V_\pi = 85$ V ($\lambda = 633$ nm). Sixteen transistors are arranged in four parallel columns consisting of four transistors in series each. The maximum collector dissipation of each transistor is 20 W. The optical frequency response of the modulator-driver system is shown in Fig. 28 (Sec-

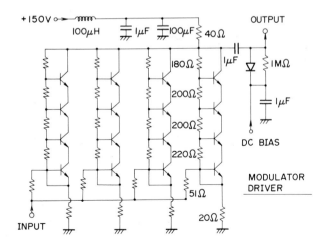

Fig. 14.   Wideband video amplifier for driving a transverse mode electrooptic modulator.

tion VIII). The response is flat from dc to 27 MHz within the fluctuation
of ± 1 dB, and falls by 3 dB at 29 MHz.

As Eq. (34) indicates, the phase retardation for an EO modulator includes a term $(n_o - n_e)(2\pi L_{y'}/\lambda)$ due to the natural birefringence, which
is temperature dependent. This term can be canceled out by arranging
two crystals in tandem with their optical axes orthogonal (Chen, 1970;
Peters, 1965). Even with this scheme, however, it is difficult to suppress
fluctuation of the residual birefringence when an intense beam of light is
passed through the modulator. This can cause a poor extinction ratio.
This is one of the reasons it is difficult to obtain good contrast ratio with
an EO modulator.

Figure 15 illustrates a method for stabilizing the modulator characteristics by means of optical feedback (Taneda *et al.*, 1973a). A pair of sensing
voltage pulses with the same magnitude but opposite polarity is inserted
into the vertical blanking of a video signal. When a light beam is modulated with this signal, the optical output of the modulator will be a pair
of pulses with the same level and polarity, provided that the bias level is
correct. However, if the output pulses are not the same, the bias voltage
will be adjusted automatically, so as to make the two pulses the same. A
similar technique has been presented elsewhere (Waksberg and Wood,
1972).

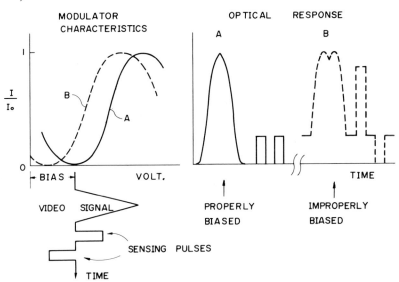

Fig. 15.   Principle of automatic bias control for an electrooptic light modulator. The
solid lines show the modulation characteristics and optical response when the modulator is properly biased. The broken lines show those when it is not properly biased.

The choice of ammonium dihydrogen phosphate (ADP) instead of KDP reduces the problems of piezoelectric resonances and unwanted noise generated by the modulator (Sliker and Burlage, 1963; Berlincourt *et al.*, 1964).

VII. Beam Deflection

A. GENERAL REMARKS

Presently available light beam deflectors are based on one of the following principles of operation: the mechanical motion of a mirror, acoustooptic diffraction, or electrooptic refraction. Among these devices, mechanical mirrors are preferable whenever possible because of the high resolution capability and achromatic property. For fast sequential scanning, multifacet rotating polygons are employed, and for slow scanning, galvanometers are available. The AO deflector, which is based on Bragg reflection of light by acoustic wavefronts, is applicable to medium resolution analog scanning and to intermediate-speed digital deflection systems. The EO deflector is suited for fast digital systems, but seems to be unlikely to find wide application in analog scanning systems. Various types of deflectors are summarized in the literature (Fowler and Schlafer, 1966; Zook, 1974; Gorog *et al.*, 1972a). For the EO deflector, reference is made to the literature (Lee and Zook, 1968; Beasley, 1971). A comprehensive survey on laser scanning systems has recently been published by Beiser (1974). This informative monograph contains a wide variety of scanning techniques, including digital systems. We treat here two important types of deflectors: the mechanical mirror deflector and the AO deflector.

The performance of a deflector is determined by the resolution capability and the scanning speed. Therefore, it is reasonable to define the figure of merit of the deflector by the maximum number of resolvable spots that can be scanned per second. For a random access deflector with a resolution of N_R resolvable spots per full deflection angle and an access time τ second to deflect the beam from one spot to another, the figure of merit equals N_R/τ. For an analog scanner which generates M_m scanning lines per second with a resolution of N_R resolvable spots per scanning line, the figure of merit equals $N_R M_m$. Since the access time of this scanner is $\tau = 1/M_m$, the expressions for the figure of merit for both the digital and analog deflectors are identical. The figure of merit is 10^6–10^7 for existing moving iron and moving coil galvanometers, about 10^8 for the AO deflectors, and 10^8–10^9 for the rotating polygonal mirror.

B. ROTATING MIRROR POLYGON

Electromechanically driven mirrors are the highest resolution deflectors known today. For fast analog scanning, a rotating polygon with polished mirror facets is most commonly used (Fig. 16). When the polygon is rotating, each facet scans a line. Hence, a polygon with n facets rotating at ω radians per second scans $n\omega/2\pi$ lines per second. The ultimate resolution capability of a polygonal scanner is determined by the maximum allowable angular velocity, which depends on the mechanical strength of the available material.

In a rotating disk made of a homogeneous isotropic material, the maximum stress occurs at the center of the disk. At the center, the tangential stress σ_θ and the radial stress σ_r are equal, and their value is

$$\sigma_r = \sigma_\theta = [(3 + \eta)/8]\rho\omega^2 R^2 \tag{36}$$

where ρ and η are the density and Poisson's ratio of the material, respectively, and R is the radius of the disk. If the maximum allowable stress is σ_m, then the maximum allowable angular velocity ω_m is

$$\omega_m = \frac{1}{R}\left(\frac{8\sigma_m}{(3 + \eta)\rho}\right)^{1/2} \tag{37}$$

The maximum number of scanning lines per second is

$$M_m = n\omega_m/2\pi = \frac{n}{2\pi R}\left(\frac{8\sigma_m}{(3 + \eta)\rho}\right)^{1/2} \tag{38}$$

Each mirror facet scans an angle Φ, which equals twice the angular motion of the mirror: $\Phi = 4\pi/n$. Therefore, the ultimate resolution given by Eq. (12) becomes

$$N_R = (8\pi^2/n^2)(R/\lambda) \qquad \text{(resolvable spots per scanning line)} \tag{39}$$

where the width of the mirror facet, $W = 2\pi R/n$, was put into Eq. (12).

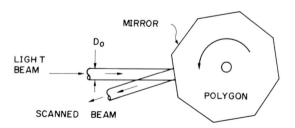

Fig. 16.   Rotating polygonal mirror scanner.

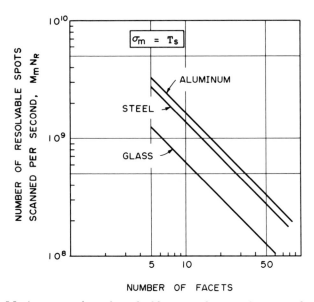

NUMBER OF FACETS

Fig. 17. Maximum number of resolvable spots that can be scanned per second, M_mN_R, versus the number of mirror facets, for a polygonal mirror scanner made of various materials. The maximum allowable stress σ_m, is equated to the ultimate tensile strength T_s. In practical cases an appropriate safety factor must be taken into account.

Multiplication of Eqs. (38) and (39) gives an expression for M_mN_R, the maximum number of resolvable spots that can be scanned by the mirror polygon per second. The result is

$$M_mN_R = \frac{4\pi}{n\lambda}\left(\frac{8\sigma_m}{(3+\eta)\rho}\right)^{1/2} \tag{40}$$

which is inversely proportional to the number of mirror facets n, but is independent of the mirror size R. Values obtained from Eq. (40) are plotted in Fig. 17.

The above discussion is based on the assumption that the incident beam is sufficiently broad to illuminate two adjacent mirror facets simultaneously. In this scheme, the facet which performs the scanning moves with its surface always fully illuminated by the beam regardless of its position during mirror rotation. Thus, the ultimate resolution capability of a rotating polygon is fully assured in this case. Furthermore, there can be no retrace time for the scanned beam to fly back from the end of a scanning line to the starting position, since the adjacent facets perform the scanning in sequence. This approach, however, wastes a large amount of beam energy. Therefore, it is applicable only to small scale devices.

Another approach is for one mirror facet to be illuminated by a small diameter beam at a conscious sacrifice of resolution capability. In the television scanning format, a finite time interval (horizontal blanking time) is allowed for the beam retrace. Therefore, the beam can be widened within this allowance. The upper limit for the beam diameter is given by

$$D_0 = \xi_H W \tag{41}$$

where ξ_H is the horizontal blanking factor and W is the facet width. If this condition is satisfied, the beam can move from one facet to the next within the blanking time. From the last expression we see that the effective aperture of the mirror facet is reduced to ξ_H times the true width in this approach.

1. *Numerical Example*

The polygon is assumed to be made of high tensile strength aluminum alloy, for which the ultimate tensile strength, T_s, is 5.0×10^9 dyne/cm², $\rho = 2.7$, and $\eta = 0.33$. We put $\sigma_m = (1/5)T_s$, where the factor 1/5 is a safety factor. Assume that $\lambda = 515$ nm, $R = 2$ cm, and $n = 25$. From Eqs. (37), (38), (39), and (40), we obtain for a fully illuminated mirror:

$$
\begin{aligned}
\omega_m/(2\pi) &= 2.4 \times 10^3 && \text{(revolutions per second)} \\
M_m &= 5.9 \times 10^4 && \text{(scanning lines per second)} \\
N_R &= 4.9 \times 10^3 && \text{(resolvable spots per scanning line)} \\
M_m N_R &= 2.9 \times 10^8 && \text{(resolvable spots per second)}
\end{aligned}
$$

For a partially illuminated mirror with a retrace time $\xi_H = 0.16$, the resolution N_R and the number of resolvable spots $M_m N_R$ is reduced to 7.9×10^2 and 4.7×10^7, respectively.

When the incident beam is narrower than twice the facet width, the beam is truncated by the mirror edge when it traverses the incident beam. Therefore, resolution and luminance obtained on the right and left peripheries of the picture are lower than those values obtained at the center. Figure 18 shows fractional decrement of beam power for a Gaussian beam of radius r_0 (at e^{-2} intensity point) truncated by a straight edge at a distance a from the axis. If we make $a = r_0$ with $r_0 = \xi_H W/2$, we find that the maximum shading occurring on the peripheries of the active picture is 2.4% (see also Dickson, 1972).

The dual beam scanning (Stone *et al.*, 1969) employs two incident beams which scan the alternate scanning lines. With this method, the beam diameter can be widened to half the facet width without a corresponding increase in retrace time. With another method (Taneda *et al.*, 1973a), the

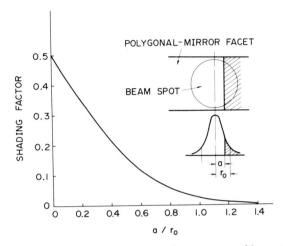

Fig. 18.   Optical power shading of a Gaussian beam truncated by a straight edge.

beam is wobbled sinusoidally to shorten the retrace time. With the sinusoidal wobbling the effective aperture of the mirror can be increased to
$0.58W$, rather than $0.16W$ for the ordinary approach in which the incident
beam is fixed.

Figure 19 illustrates a driving mechanism for a high-speed rotating
mirror scanner used in the recently developed 1125-scanning-line laser
color-TV display system (see Section VIII). A 25-facet polygon made of

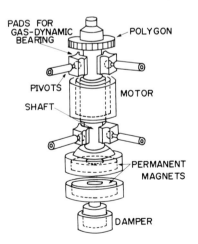

Fig. 19.   High-speed rotating polygon supported by magnetic thrust bearing and
gas-dynamic radial bearings.

aluminum alloy is mounted on a rotor that is driven by a high-speed synchronous motor at 81,000 rpm. The shaft of the rotor is supported by a magnetic bearing in the upward direction and by gas-dynamic bearings in the radial direction. The magnetic bearing consists of a pair of ring-shaped permanent magnets made of strontium ferrite, which are 8 mm thick, 17 mm in inner diameter, and 36 mm in outer diameter. The two magnets repel each other, thereby lifting the $420g$ rotor 6 mm. The rotating magnet is embedded in a titanium alloy cup to strengthen it against the centrifugal force. Since the stator magnet is allowed to slide in the vertical direction, vertical oscillation of the rotor is absorbed by the oil damper.

The gas-dynamic bearing consists of movable pads made of hardened steel. When the shaft is rotating, air is forced into the gap between the shaft and the pads, which prevents the shaft from direct contact with the pads. An external supply of compressed air is not needed after the motor has started rotating.

The 25-facet polygonal mirror is 40 mm in diameter and 5 mm thick. Each facet is a 5×5-mm square. The mirror facet is nickel plated, polished, and then coated with evaporated aluminum and silicon monoxide.

The high-frequency power source driving the motor is phase-locked to the horizontal synchronizing signal by means of a voltage-controlled oscillator (VCO). The phase signal of the motor can be obtained from a magnetic tachometer, or by using a He–Ne laser beam reflected from the facets. Measured jitter of this deflector was ±20 ns.

C. RASTER IRREGULARITY COMPENSATION

With a special machining technique the polygon can be machined to within angular tolerance of 10 arc sec. However, even higher precision is required in high-quality display systems, in which a very small amount of residual error in the parallelism between the mirror facet and the rotating axis can cause a noticeable amount of raster irregularity in the generated image. Previously, this irregularity was compensated by means of an EO beam deflector (Taneda *et al.*, 1973a). More recently (Taneda *et al.*, 1973b), a pair of cylindrical lenses was used to compensate for the raster irregularity. In the top view of Fig. 20, the presence of a pair of cylindrical lenses CL_1 and CL_2 can be ignored, since these lenses have no effect on the horizontal deflection of the beam. The relay lenses L_1 and L_2 are used to reduce the size of the vertical scanning mirror; i.e., the nutating bundle of light converges on the galvanometer mirror irrespective of the prior horizontal scanning. By the adjustment of the relay lenses the beam can be focused to form a beam waist as shown in Fig. 3.

In the side view of Fig. 20, the dashed lines show the beam trace when

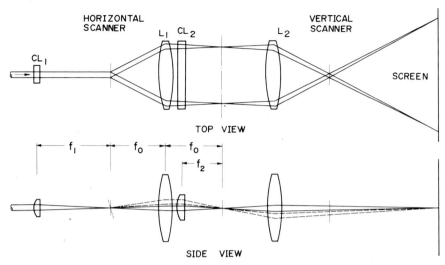

Fig. 20. Raster irregularity compensation by cylindrical lenses CL_1 and CL_2. The broken lines show the optical path when the mirror facet of the horizontal scanner is tilted at a small angle (as shown in the side view of the figure).

the horizontal scanning mirror is tilted at a small angle from the proper position. This beam can be directed to its proper position on the screen by the cylindrical lens CL_2, which forms, in combination with L_1 and L_2, the image of the horizontal scanning mirror on the viewing screen. Deformation of the beam cross section introduced by the cylindrical lens can be corrected by placing another cylindrical lens CL_1 as shown in the figure. The focal lengths of the lenses CL_1, CL_2, and L_1 must satisfy the following condition:

$$f_1 f_2 = f_0^2 \qquad (42)$$

where f_0 is the focal length of L_1, and f_1 and f_2 are those of CL_1 and CL_2. Figure 21 shows the effect of introducing the raster irregularity compensator. Results with and without the cylindrical lenses are shown.

D. GALVANOMETER

The galvanometer deflector is a good device for high resolution, slow access beam deflection. This device consists of a small mirror mounted on the electromagnetically driven armature or coil of a galvanometer. Figure 22 shows the magnetic circuit of a moving-iron galvanometer (Brosens and Grenda, 1974). A rotating armature forms four air gaps with corresponding stationary pole pieces, and a balanced bias magnetic flux is produced at each gap by a pair of permanent magnets. A pair of coils applies

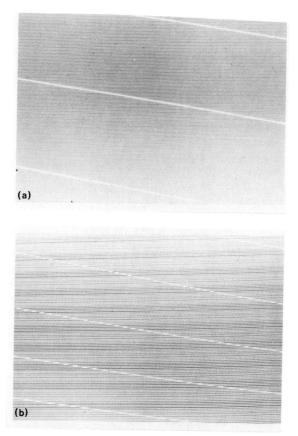

Fig. 21. Photograph of the raster generated with (a) and without (b) the raster irregularity compensator. A uniform raster is obtained with the compensator.

additional magnetic flux at the air gaps in such a way as to reinforce the bias flux across one pair of diametrically opposed air gaps and to reduce it across the other. As a result, the armature experiences an electromagnetic torque, which causes it to align itself with the air gaps of the strongest flux. Through careful design and adjustment of the bias flux produced by the permanent magnets, it is possible to produce a torque which is linearly proportional to coil current and independent of armature rotation over an angle of up to 30°. Therefore, if the armature is fastened to a torsional spring in the same manner as the moving coil of d'Arsonval movement, armature deflection proportional to the coil current can be obtained.

The limiting resolution of a galvanometer is determined by the maximum mirror tilt angle and the mirror aperture [see Eq. (12)]. Although the resolution is proportional to the mirror aperture, inertia also increases with mirror size. The maximum speed required to deflect a beam from one position to another (access time) is determined by the natural resonant frequency of the galvanometer. By using a precisely timed two-step waveform, a galvanometer can be driven with an access time of about half the resonant period with no overshoot (Brosens, 1971; Berg et al., 1974). The resonant radian frequency ω_0 is given by

$$\omega_0^2 = K/(J_m + J_a) \qquad (43)$$

where K is the torsion constant (ratio of torque to tilt angle), and J_m and J_a are the moments of inertia of the mirror and the armature. For a square mirror of aperture width W and mass m rotating about its center of gravity, $J_m = (\frac{1}{12})mW^2$. For a circular mirror of diameter W, $J_m = (\frac{1}{16})mW^2$.

A square glass mirror with 5-mm sides and 0.5-mm thickness has inertia of $J_m = 6 \times 10^{-4}$ g cm^2. When loaded with this mirror a galvanometer with $J_a = 1 \times 10^{-2}$ g cm^2 and $K = 1 \times 10^6$ dyne cm (values for a typical moving iron galvanometer) will oscillate at the resonant frequency of 1.5 kHz. If the mirror is driven to the rotation angle of 8 deg peak to peak, the resolution is 900 spots per scan ($\lambda = 0.5$ μ). The figure of merit of this scanner is 3×10^6 spots per second.

Performance characteristics of commercially available galvanometers* are summarized in the literature (Beiser, 1974; Zook, 1974). Moving iron

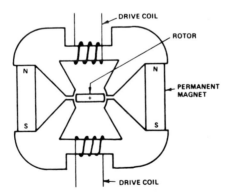

Fig. 22. Magnetic circuit of a moving-iron galvanometer. (From Brosens and Grenda, 1974. Reproduced with permission of the author and the publisher, Electro-Optical Systems Design).

* Major suppliers: General Scanning Inc., 150 Coolidge Avenue, Watertown, Maine 02172, and Honeywell Test Instruments Division, Denver, Colorado 80217.

galvanometers have an unloaded resonant frequency of 0.13–3.2 kHz, and can be loaded with a mirror with 5- to 25-mm aperture. Moving coil galvanometers are loaded with a miniature mirror (aperture less than 1 mm), and oscillate at a resonant frequency of up to 22 kHz. The figure of merit of existing moving iron and moving coil galvanometers is between 10^6 and 10^7 spots per second.

Operation of a galvanometer in general is quite stable without feedback control. However, scanning errors can be further reduced by position sensing to generate an error signal. This signal is fed back into the coil drive circuits (Grenda and Brosens, 1974).

E. AO DEFLECTION

In the acoustooptic Bragg reflection, the externally observed angular separation between the zero and the minus first order beams is, from Eq. (26), $2\theta_B = K/k = \lambda/\Lambda = \lambda f/V$, where f is the sound frequency and V is the propagating velocity (see Fig. 23). If the frequency is increased by an amount Δf, the incremental deflection angle for the minus first-order beam is

$$\Delta\theta = (\lambda/V)\Delta f \tag{44}$$

Therefore, the direction of the diffracted beam can be steered by changing the acoustic drive frequency. If the aperture width of the acoustic cell is W, then the limiting resolution of the AO deflector is

$$N_R = \Delta\theta W/\lambda = (W/V)\Delta f = \tau\Delta f \tag{45}$$

where τ is the transit time of the sound wave across the aperture width. Since τf is the number of sound waves present within the aperture, $\tau\Delta f$ is the number of waves added when the acoustic frequency is changed by Δf.

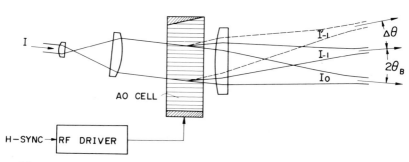

Fig, 23. Acoustooptic (AO) beam deflection. Angular separation between the zero and the minus first-order beams equals twice the Bragg angle $2\theta_B$. The broken lines show the beam position when the acoustic frequency is increased by Δf.

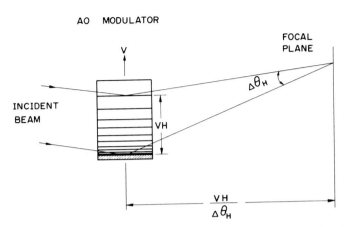

Fig. 24. Cylindrical lens effect of a linearly frequency-swept acoustooptic (AO) modulator (Korpel et al., 1966).

In a digital deflection system in which the beam is steered from one position to another in a random access format, the switching time is limited by the acoustic transit time τ. For a fixed frequency swing Δf, the resolution can be increased only at the sacrifice of the access time. Assuming that $N_R = 1000$ resolvable spots and $\Delta f = 100$ MHz, then we have $\tau = 10$ μs. If this deflector is made of lead molybdate ($V = 3.6$ mm/μs), then $W = V\tau = 36$ mm.

When Eq. (45) is applied to a television display system with a minimum picture element of 0.15 μs and a resolution of about 200 spots per scanning line, we find that a frequency swing of $\Delta f \approx 1$ GHz is required. This exceeds the capabilities of present technology.

However, a television display with linear analog scanning is a special case (Korpel et al., 1966; Gerig and Montague, 1964). In this case the sound frequency is modulated linearly with respect to time t. At the same time, the sound wave which deflects the light beam by $\Delta\theta$ travels in the direction y at a velocity V (see Fig. 24). The deflection angle as a function of y and t is

$$\Delta\theta(y, t) = (\Delta\theta_H/H)(t - y/V) \qquad (46)$$

where $\Delta\theta_H$ is the total angle of deflection during the horizontal scanning time H. The spatial distribution of deflection angle at any given instant is as follows:

$$(\partial/\partial y)\Delta\theta = -\Delta\theta_H/(HV). \qquad (47)$$

which is independent of t and y. Since the deflection angle changes at a uniform rate along the y axis, the AO modulator operated in the linear

FM mode acts like a cylindrical lens with a focal length of $VH/\Delta\theta_{\rm H}$. Since this effect can be compensated for by introducing a conventional cylindrical lens immediately preceding or immediately following the deflector, the acoustic transit time τ need not be made very small in the analog scanning system. In actual display systems the upper limitation on τ is set by the horizontal retrace time $\tau_{\rm max} = \xi_{\rm H}H$. This is about 10 μs for the standard 525-line format.

The condition of Bragg reflection holds strictly only for a specific sound frequency, for which the incident and diffracted beams are symmetrical with respect to the acoustic wavefronts. Since the position of the incident beam is fixed in practical devices, the acoustic wavefronts must be rotated to follow the Bragg angle when the acoustic frequency is changed. This can be achieved (as shown in Fig. 25) by means of attaching a phased array of transducer elements to a staircase. The adjacent transducers are driven in opposite phases. Each step is $\frac{1}{2}\Lambda_o$ high and $\Lambda_o{}^2/\bar\lambda$ wide, where Λ_o is the acoustic wavelength at the center of the frequency swing, and $\bar\lambda$ is the wavelength of light in the medium (Korpel *et al.*, 1966). In another approach (Coquin *et al.*, 1970; Alphonse, 1974) a multielement transducer array is formed on the flat surface of the medium, and the transducers are driven by the input signals whose phases are shifted in sequence.

In the acoustooptic effects which have been considered in this chapter, the diffraction process did not change the polarization of the light and the birefringence of the medium was not included. In isotropic media, the Bragg diffraction geometry is simplified by the basic relation $|\,\mathbf{k}\,| \simeq |\,\mathbf{k}_{-1}\,|$, where $\mathbf{k}$ and $\mathbf{k}_{-1}$ are the incident and diffracted optical wavevectors, re-

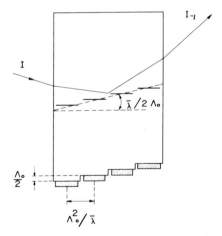

Fig. 25.   Phased array acoustooptic deflector (Korpel *et al.*, 1966).

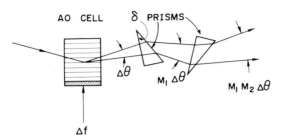

Fig. 26. Deflection angle multiplier which provides correction for the chromatic aberration of an acoustooptic (AO) device (Watson and Korpel, 1969; Hrbek et al., 1970, 1971).

spectively. In an anisotropic medium, the Bragg relation requires modification (Dixon, 1967): the optical birefringence usually requires $|\mathbf{k}| \neq |\mathbf{k}_{-1}|$ if the diffracted optical polarization differs from the incident. Anisotropy modifies the directions into which the induced polarization may radiate and thereby changes the Bragg diffraction geometry. Anisotropic AO deflectors have been constructed using paratellurite (TeO_2) (Uchida and Ohmachi, 1970; Warner et al., 1972) and other materials (Lean et al., 1967). A recently developed laser display system (Gorog et al., 1972b) employs an anisotropic TeO_2 deflector. The acoustooptic interaction length required is only 0.225 cm, and the acoustic frequency sweep band is 65 MHz, extending from 30 to 95 MHz. For the optical aperture of 0.5 cm, the diffraction-limited resolution was 524 spots.

In acoustooptic interaction, the diffraction angle is dependent on the optical wavelength. Therefore, in multicolor display systems, the deflection angle of two of the primaries must be magnified or demagnified to match the third. Figure 26 shows a scanning angle multiplier that provides the correction (Watson and Korpel, 1969; Hrbek et al., 1970, 1971). If a prism with apex angle δ and refractive index n is oriented as shown in this figure, the deflection angle is multiplied by a factor

$$M(\phi) = d\theta/d\phi$$

$$= \frac{(\sin \delta \sin \phi \cos \phi)/n \cos [\sin^{-1}(\sin \delta/n)] + \cos \phi \cos \delta}{[1 - n^2 \sin^2 (\delta - \sin^{-1} (\sin\phi/n))]^{1/2}}$$

where ϕ is the incident angle of the light beam measured from the normal to the front surface of the prism, and θ is the exit angle measured on the rear surface. For normal incidence ($\phi = 0$) the multiplication factor reduces to $\cos \delta/(1 - n^2 \sin^2 \delta)^{1/2}$. The prism can be rotated to make adjustment of the magnification factor. For a fixed prism position, the magnification changes as the light is deflected. This would result in misregistration of the blue and green images with the red. This aberration can be canceled by using two prisms in cascade, as shown in Fig. 26.

VIII. An 1125-Scanning-Line Laser Color-TV Display

Recently, a high resolution laser color-TV display system was developed through collaboration of the NHK (Japan Broadcasting Corporation) Technical Research Laboratories and the Central Research Laboratory of Hitachi, Ltd. (Taneda *et al.*, 1973b; Yamamoto, 1975a). To take advantage of the high-resolution capability of laser displays, this system employs an unconventional scanning format of 1125 scanning lines per frame. The vertical scanning rate is 30 frames per second (2 to 1 interlace).

A system block diagram is shown in Fig. 27. An argon ion laser with a CW multimode output of 8 W (all lines) generates a blue primary at 476.5 nm (1 W) and a green primary at 514.5 nm (3 W). These two wavelengths are separated from each other by means of a dispersive prism. The intensity of the green primary is adjusted with an optical attenuator to obtain balance with the blue intensity. A krypton ion laser with optical power output of 2 W at 647.1 nm provides a red primary. The three primary beams are passed through respective modulators, and are then combined onto a common axis by means of dichroic mirrors. The rotating polygon for horizontal scanning employs magnetic and gas-dynamic bearings. This scanner has been described in Section VII, B. A He–Ne laser is used to obtain phase signals from the rotating polygon. The vertical scanning is accomplished by a galvanometer (General Scanning, Inc., Model G-108). The raster irregularity compensator described in Section VII, C is incorporated in the scanning system.

The video signal is generated by a three-vidicon, 1125-scanning-line camera. The RGB outputs are processed in a contour compensator. Since this camera is normally adapted to the NTSC primary colors, mismatching with the laser primaries is corrected in a cross-matrixing circuit. Gamma of the total system composed of the camera, the video signal processor, and the optical modulation system is linearized by a nonlinear preamplifier.

The system can be operated with either EO or AO modulators. The EO modulators (ISOMET, Inc., Model TFM-535) are driven by the wideband video amplifier described in Section VI. Frequency response of the system is shown in Fig. 28. The abscissa is horizontal resolution expressed in TV-lines per picture height, and modulation frequency in megahertz. The limiting resolution is 1000 TV lines per picture height. The vertical resolution is 700 TV lines per picture height. A contrast ratio of 30 to 1 can be

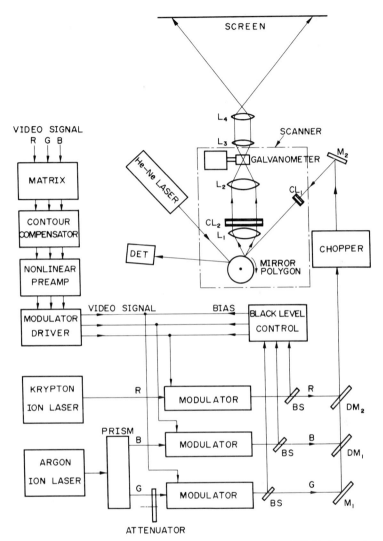

Fig. 27. Block diagram of an 1125-scanning-line laser color-TV display system. BS: Beam splitter. M_1, M_2: Reflecting mirrors. DM_1, DM_2: Dichroic mirrors. DET: Photodetector. L_3, L_4: Projection-angle adjustment lenses.

obtained when the automatic bias control system described in Section VI is incorporated. The chopper shown in Fig. 27 is needed to blank out the sensing pulses (see Section VI), with which operating characteristics of the EO modulator are monitored.

The AO modulator used with this system is the model DLM-20 modu-

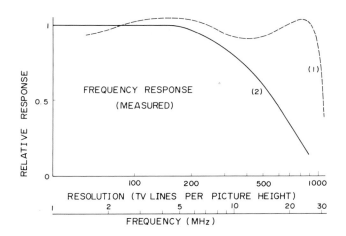

Fig. 28.　Frequency response of an 1125-scanning-line laser color-TV display system. Curve (1): Response of the optical modulation system composed of an electrooptic light modulator and a video amplifier. Curve (2): Response of the total system.

lator and driver supplied from Datalight Inc. With the AO modulators an excellent contrast ratio (better than 50 to 1) can be obtained without servo control system, in contrast to the EO system, which is very sensitive to temperature variations. The AO modulation system is further simplified by the low power requirement of the AO modulator. With AO modulators the horizontal resolution is reduced to 600 TV lines per picture height, but the vertical resolution is not altered.

Figure 29 shows the display system in operation. The optical system is assembled on a 90 × 130 cm steel table. When a picture of 3-m² active area is projected onto a screen of gain 4, the highlight luminance is 11 fL. Luminous flux generated by the lasers is 550 lm and the optical transmission coefficient of the system is 0.2.

Figure 30 is a photograph of the resolution test pattern and Fig. 31 shows the test subject, both taken from the projection screen. Although the photograph of Fig. 31 is reproduced here in only black and white, in the full-color image generated on the viewing screen the skin tones and white are faithfully reproduced and the color gamut obtained is quite good. Despite the fact that the wavelength of the blue primary is longer than desired, the difference can hardly be recognized unless the laser image is closely compared with the CRT monitor.

This system has been used at the NHK Technical Research Laboratories for the study of high definition television.

Fig. 29.   An 1125-scanning-line laser color-TV system in operation. The projected image is 1.5 m high and 2 m wide.

IX. Applications of Scanned Beam Techniques

In the previous section we demonstrated the capability of the laser to generate high-quality images on a large projection screen. However, large screen displays are not the only field in which the scanned beam technique may find application. As the capability of lasers to generate high-quality images become more widely understood, additional applications of the laser beam will be developed. A great deal of work is being done on the use of laser beams for image scanning and recording, pattern generation, etc. In a laser scanner, the laser beam is modulated by the transparency of the film on which the beam is scanned. The modulated light is collected on a photodetector to convert the image to a video signal. In a recorder, the incoming video signal is used to modulate the laser beam, and the modulated light is recorded on unprocessed film to produce a picture. The laser pattern generator is used, for example, as a mask-making device in the field of thin film technology. These applications are briefly surveyed in this section.

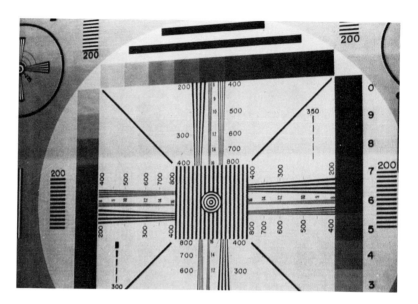

Fig. 30. Photograph of the central portion of an image projected on the screen.

A. IMAGE SCANNER AND RECORDER

1. *Color Television Film Recording*

An application for which scanned beam technology is well suited is the direct recording of color television on color motion picture film (Schlafer and Stone, 1969; Beiser *et al.*, 1971). This technique is preferred for subse-

Fig. 31. Photograph of a test subject projected on the screen.

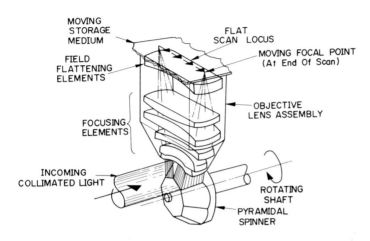

MOVING STORAGE MEDIUM

FIELD FLATTENING ELEMENTS

FLAT SCAN LOCUS

MOVING FOCAL POINT (At End Of Scan)

OBJECTIVE LENS ASSEMBLY

FOCUSING ELEMENTS

INCOMING COLLIMATED LIGHT

ROTATING SHAFT

PYRAMIDAL SPINNER

Fig. 32. Typical high resolution image recorder. Incoming collimated light is intercepted by two facets and redirected through objective lens to form a moving focal point along a flat locus. (From Beiser, 1972. Reproduced with permission of the author and the publisher, Industrial and Scientific Conference Management, Inc.).

quent presentation or distribution of television programs because of the cost and complexity of high-quality color video tape recording and playback equipment. Figure 32 shows an optical system for film recording using a pyramidal scanner and a flat field objective lens assembly (Beiser, 1972). White highlight exposure on Kodak 7241 16-mm color film requires an optical power level of 0.08 μW for 647.1 nm to 1.4 μW for 514.5 nm (Schlafer and Stone, 1969; Fowler et al., 1971). With these low power requirements, a single multicolor laser such as a krypton ion laser or a helium–selenium laser (Silfvast, 1970) can be utilized as the light source. Figure 33 shows relative spectral sensitivity of the color layers in a typical color-reversal film with wavelengths of the Kr^+ and $HeSe^+$ laser (Schlafer and Stone, 1969; Fowler et al., 1971). The laser film recording eliminates color registration problems. Therefore, it provides higher resolution compared with the present method in which a standard shadow-mask CRT is photographed.

2. High Resolution Image Scanners and Recorders

Considerable effort is being directed to development of a system for accurately reading and recording documents and images. A high-resolution laser image scanner and recorder developed by RCA Corporation (Dobbins, 1972; Kenville, 1971; Woywood, 1972) is capable of resolving more than 20,000 picture elements at 50% MTF (modulation transfer function) with a geometric distortion of one part in 20,000. The 5-inch-wide film is curved

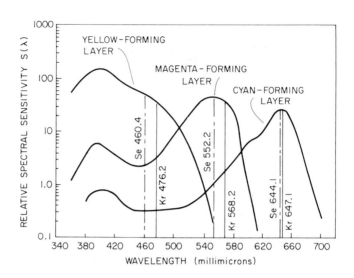

Fig. 33. Relative spectral sensitivity of the three layers in a typical color-reversal film. The wavelengths of the Kr⁺ and He–Se⁺ laser are also shown. (From Fowler *et al.*, 1971. Reproduced with permission of the authors and the publisher, The Institute of Electrical and Electronics Engineers, Inc.).

concentrically with a six-faceted pyramidal scanner. The film is supported on both sides by a hydrostatic air bearing within 200 microinches of the focus. The vertical scanning is accomplished by motion of the film. The light source is an air-cooled argon ion laser capable of 45 mW output. The inherent spot position error due to mirror facet tolerance ($\frac{1}{4}$ of a spot) and scan motor jitter (1.2 spots) is removed by a variable delay line inserted into the video path. Multispectral image data can be stored in a single frame on black-and-white film with a diffraction grating technique.

The image recorder developed at Itek Corporation employs a high-precision pyramidal mirror scanner made of beryllium. The maximum number of picture elements per line is 36,800 over a format of 228.6 mm (Gramenopoulos and Hartfield, 1972).

An experimental film recorder being developed for the Earth Resources Satellite Program converts three-color input signals into an image recorded on color film (Montuori *et al.*, 1973). The scan area is 9.5 × 9 inches, and the system response is 70% at 20 cycles/mm.

Analysis of available high-resolution laser recording technique was made and a new approach to beam scanning was described (Yellin, 1972).

A laser image processing scanner shown in Fig. 34 was developed by CBS Laboratories (CBS Laboratories Technical Data Sheet). A photographic transparency is scanned with a 1.25- to 10-μm diameter laser beam. The film is mounted on a precision quartz cylinder with a photo-detector inside the cylinder. The scanning is accomplished by rotating the cylinder and driving an optical carriage in the axial direction. The analog signal derived from the film is digitized and processed by computer. The processed signal then modulates the laser beam which records the recon-structed and processed image on an adjacent portion of the same rotating film. A single laser is used for both reading and writing.

A high-resolution microfilm recording system developed in Bell Labora-

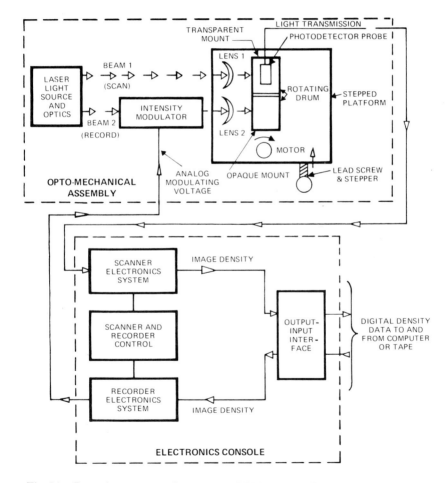

Fig. 34. Laser image processing scanner (LIPS). (Reproduced with permission of CBS Laboratories, Inc.).

tories (Cheng and Miller, 1973) is based on a Nd-YAG laser operated in a cavity-dumped mode with 100-mW TEM_{00} average power. [For the cavity-dumped mode of operation, reference is made to the literature (Geusic *et al.*, 1970)]. The resulting 0.3-MHz pulses are modulated by a TeO_2 acoustooptic modulator and deflected by a gimballed galvanometer mechanism. The modulated pulses make holes ranging from 0 to 10 μm in diameter by melting bismuth film coated on Mylar film. The picture consists of a two-dimensional array of these varying-sized holes. The resolution is 1500 picture elements per line and 2000 lines per frame. The typical horizontal scanning rate is 160 Hz (500 Hz maximum) and the vertical scanning period is 12 s (4 s minimum).

3. Facsimile

An economical laser facsimile system has been developed which is compatible with existing systems but has the advantage of superior picture quality (Schreiber, 1973). The system is based on a 1-mW He–Ne laser, which is internally modulated by discharge current, a galvanometer-driven mirror for horizontal scanning, continuous paper motion for vertical scanning, and heat processed dry silver paper as the recording medium. The system, originally designed for newspaper use, handles paper 11 inches in width and of any length. The scanning density is 100 lines per inch at 100 lines per minute. This is compatible with the transmission bandwidth of present telephone lines. Sensitivity of 3M Type 7771 dry silver paper is 200 erg/cm^2 for full blackening.

The graphics transmission machine developed at Bell Laboratories (Berg *et al.*, 1974) employs a 10-mW He–Cd laser which simultaneously scans an $8\frac{1}{2} \times 11$ inch ($21\frac{1}{2} \times 28$ cm) original page and an image recorded on vesicular diazo heat developed film. The image is reduced to $\frac{1}{24}$ of the original size, or 9 $\times$ 11.6 mm. The reading and writing beams are obtained from the same beam by the use of a beam splitter. An acoustooptic modulator is used to modulate the beam with the video signal, which is picked up from the original document by suitably arranged photodetectors. Raster scanning is accomplished by a vertical scan galvanometer operating at four seconds per frame and a horizontal scan galvanometer operating at 500 Hz (1.4 ms of active scan and 0.6 ms for retrace).

The microfilm supplied from Kalvar Corporation requires about 20 mJ/cm^2 for proper exposure to full working density. The raster for one frame contains 2000 lines with equivalent resolution per horizontal scan. Contrast ratios of up to 40 to 1 can be achieved. The video bandwidth utilized for transmission is 500 kHz.

B. LASER PATTERN GENERATION

A scanned beam system has been developed that produces a thin film mask of desired pattern under computer control (Raamot and Zaleckas, 1974). According to these authors the exposure mask used in the thin film technology defines, in conjunction with exposure and etching procedures, the features of the final mask. The present method of producing masks consists of generating an oversized emulsion master that is then photographically reduced to form the final mask. In many cases, the emulsion copy is converted to a metal mask by additional subsequent steps. The laser pattern generator (developed by Raamot and Zaleckas) generates the required mask directly in a fast, one-step process, by the selective machining of a thin film metal blank such as chromium, tantalum, or iron oxide deposited on a transparent glass substrate. By the use of galvanometers for x-y beam deflection, the system can be operated either in a random access format under computer control, or in a raster scanning format to produce a copy of the master transparency interrogation. A Nd:YAG laser is operated in a 50-kHz Q-switch mode with average power output of 20 W. A He–Ne laser with a 1.15-μm output beam is used as the reference for the beam position monitor.

A maximum field size of 100 × 100 mm and a machined spot size of 25 μm result in a resolution of 4000 × 4000 spots. The line drawing rate is 500 mm/sec. Position accuracy is ±5 μm (dynamic) and positioning and acceleration time is typically 1.5 ms.

C. OTHER APPROACHES AND MISCELLANEOUS USES

The direct-view laser display in which the visible laser beam supplies its own light to the screen has been the focal point in this chapter. Although this approach eliminates limitations on screen size and material, the laser has low luminous efficiency as a light source. Another approach is the laser-beam-addressed light valve, in which the laser beam provides local control of an optical parameter of some material (i.e., reflectivity or transmissivity) and a separate conventional lamp provides the light to the screen. An example is the laser-addressed light valve using liquid crystals (Maydan, 1973; Kahn, 1973; Melchior et al., 1972). Cholesteric or smectic liquid crystal is sandwiched between two substrates such as glass or fused silica plates. The substrates are coated with electrodes which are absorbing in the infrared, and transparent in the visible. A focused infrared laser beam is absorbed by the electrodes and locally heats the liquid crystal from its well-ordered state into a disordered isotropic phase. When the liquid crystal cools off, the state freezes and scatters light. The image can be viewed by a schlieren projection system, and is erased by applying a voltage across the cell. Typically a Nd:YAG laser with 40-mW output records graphic information at a writing speed of 10^5 spots per second with a contrast ratio of better than 10 to 1. Resolution of over 2000×2000 pictorial elements can be obtained (Maydan, 1973).

It was shown that monochromatic blue or ultraviolet light can be converted into virtually any visible color including white by properly coating a viewing screen with existing organic and inorganic phosphors (Van Uitert *et al.*, 1971; Pinnow *et al.*, 1971). As an additional benefit of this conversion the speckle noise associated with direct viewing of scattered light is eliminated, because the converted light is incoherent. A black and white display can be achieved with a blue argon ion laser beam at 488.0 nm when combined with an appropriate blend of blue to red conversion phosphor and direct scattering material such as powdered MgO.

The scan laser is an image generator in which in-cavity control of the oscillation provides a two-dimensional image directly from the laser source. Scan lasers have been built using Nd:YAG (Chang *et al.*, 1970; Dakss and Powell, 1968) and semiconductors (Packard *et al.*, 1971; Kotovshchikov *et al.*, 1974).

Scanned laser beam technique has been applied for testing IC's (McMahon, 1971; NASA, 1973). Laser scanning microscopes have been described by many authors (Davidovits and Egger, 1971; Sawatari, 1973; Sommargren and Thompson, 1973; Black *et al.*, 1972; DiStefano and Viggiano, 1974; Williams and Woods, 1972).

X. Concluding Remarks

Good image quality has been demonstrated with large screen laser display systems, and various applications are being sought for image pickup, recording, pattern generation, etc.

With large screen displays, a serious limitation is, on occasion, imposed on its practical utility by the low conversion efficiency of gas lasers. One of the most promising candidates which may provide visible coherent light with high conversion efficiency is the Nd:YAG laser. The infrared output can be converted into visible light of any desired wavelength through a parametric up-conversion technique in combination with tunable dye lasers. A pulsed mode of operation is preferred in these wavelength-conversion processes, since conversion efficiency in nonlinear optical effects increases with increasing peak power, and fluorescent dyes have a short lifetime at upper laser transition levels. A method is presented in the Appendix in which a stationary image can be reproduced from temporal video signals using a repetitively pulsed laser.

The characteristic speckle pattern produced by coherent light in a laser display annoys many observers. Potential eye safety hazards due to laser beams is another problem of concern. These problems may be solved by forming the image on an intermediate, moving, diffusing surface such as ground glass (Baker, 1968). A conventional projection lens may then be used to project the final image on a viewing screen.

Appendix: Acoustooptic Imaging with a Pulsed Laser

Conventional laser display systems utilize a continuous-wave light beam which is temporally modulated by video signals. In this Appendix, another scheme for the laser display using a pulsed laser is presented (Yamamoto, 1974). With this scheme, an image is generated by means of an acoustooptic spatial modulator illuminated by a pulse-operated laser source. The use of the acoustooptic spatial modulator is reminiscent of the old Scophony television projector and the more recent Zenith approach; both systems, however, are based on a CW light source, as described in Sections II and VI.

The pulsed-laser display system is shown in Fig. 35. In this figure an expanded beam of light is incident on an acoustooptic modulator, which has an extended aperture along the direction of sound wave propagation. The Bragg-diffracted beam carries an image of the acoustic field pattern, which is spatially modulated by the video signal. The distance of acoustic

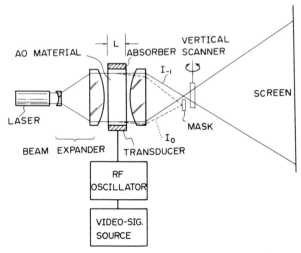

Fig. 35. Schematic diagram showing the principle of acoustooptic image generation using a pulsed laser.

wave propagation corresponds to the length of a horizontal scanning line. The linear image moves at a velocity corresponding to the sound velocity. However, it can be immobilized through stroboscopic illumination from a pulsed light source, which is repetitively operated in synchronization with the horizontal scanning. The vertical scanning is achieved in the usual manner using a galvanometer.

In an experiment a commercially available AO modulator (DLM-20, supplied by Datalight, Inc.) was used as the spatial modulator. Although the acoustic propagation distance of this modulator was only 2 cm, or one-tenth of the distance required to hold the full length of a horizontal scanning line of the standard system, it was sufficient for an experimental demonstration of the operating principle. The AO modulator was made of a material whose sound velocity was about 3.6 mm/μs. The piezoelectric transducer attached to one end of the AO material had a width of about 0.5 mm in the direction perpendicular to both the optical and the acoustic wave propagation. The sound frequency was 80 MHz.

In the experiment, the 80-MHz sound wave was modulated by 1-MHz square pulses. In synchronization with this modulating signal, a Nd:YAG laser was repetitively Q-switched at 300 Hz. The infrared output of this laser was converted to 0.53-μm green light by means of an intracavity frequency doubling process using lithium iodate as the nonlinear optical element. The green light beam was passed through a beam expander to obtain a broad beam, which illuminated the AO modulator. The image was projected with the minus first order beam on a ground glass plate.

Figure 36 shows an example of the pattern obtained in this experiment (Yamamoto and Kohmoto, unpublished observations). Due to the small width of the transducer ($\sim$0.5 mm), the image exhibits a typical narrow-slit diffraction pattern. An acoustic field with smaller divergence could

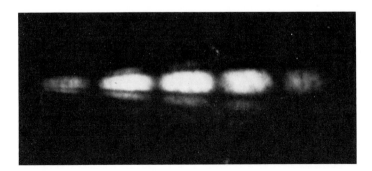

Fig. 36.   Image of an acoustooptic modulator generated by a repetitively Q-switched laser. Square-wave modulated acoustic waves are propagating from the left to the right.

have been obtained by the use of a modulator with a wider transducer. The image resolution is limited in this experiment by the duration of Q-switched pulses, which was 0.2 μs.

It is worth noting that a CW-pumped, repetitively Q-switched Nd:YAG laser has an output power capability which is, on the average, comparable with the power obtained in the CW oscillation mode; this is true provided that the repetition period is made short enough compared with the lifetime of the upper laser transition level, which is 230 μs (Chesler et al., 1970). Since the horizontal scanning time for the standard television format is shorter than this time, the Nd:YAG laser can be operated in the pulse mode with an average power output comparable with that obtained from the same laser when it is operated in the CW oscillation mode. This means that the peak power of the light pulses obtained from the Q-switched laser increases by a factor equal to the ratio of the repetition period to the pulse duration, when compared with the CW output. On the contrary, conventional flash lamps are unlikely to satisfy the requirements imposed on the light source of this system, since they have a relatively long pulse duration, low repetition rate, and poor monochromaticity.

The length of the AO modulator required is determined by the acoustic propagation distance $(1 - \xi_H)VH$, where V is the sound velocity, H is the horizontal scanning time, and ξ_H is the fractional blanking time. In the standard broadcast system, the required length is about 20 cm for lead molybdate and 14 cm for α-iodic acid. Attenuation of sound waves may lead to nonuniform distribution of acoustic power along the direction of sound propagation. This can be compensated for by superposition of a monotonically decreasing waveform on the incoming video signal. The expression for the waveform is $A(t) = \exp[\alpha V(H - t)]$, where an exponential attenuation with a decay constant α (cm^{-1}) has been assumed for the propagating sound waves.

The method presented here will, in principle, find applications for any sort of imaging systems in which a temporally modulated electric signal is to be converted into a spatial pattern of intelligence.*

Acknowledgments

The authors wish to express their sincere gratitude to Dr. B. Kazan of the Xerox Corporation, who created this opportunity for us, in addition to providing invaluable

---

* Optical information processing is one of these applications. Previous work on the use of the acoustooptic modulator in image processing has been based on the generation of optical pulses by extracavity modulation of a CW laser (Ueda et al., 1968). This method, however, wastes a large amount of optical energy. More recent work (Roberts et al., 1974) is based on an argon ion laser which is operated in a mode-locked, cavity-dumped mode generating high repetition rate light pulses.

guidance during the course of this work. Dr. L. Beiser of CBS Laboratories has kindly supplied them with literature related to laser scanning techniques. Section VIII has been written on the basis of the work performed in collaboration with colleagues in the NHK (Japan Broadcasting Corporation) Technical Research Laboratories and the Central Research Laboratory of Hitachi, Ltd. Thanks are also due to Mr. F. J. Kurdyla and Dr. A. Fukuhara for reviewing the manuscript.

References

Ahmed, S. A., and Campillo, A. J. (1969). He-Ne-Cd laser with two color output, *Proc. IEEE* **57**(11), 2084–2085.

Ahmed, S. A., and Keeffe, W. M. (1974). Parametric and discharge studies of three-color gas-mix ion lasers, *J. Appl. Phys.* **45**(1), 182–186.

Alphonse, G. A. (1974). Broad-band, acousto-optic deflectors using sonic grating for first order beam steering, *RCA Rev.* **33**(3), 543–594.

Alsabrook, C. M. (1966). A multicolor laser display, *Proc. Annu. Aerosp. Electron. Conf., 18th*, Dayton, Ohio, pp. 325–331.

Baker, C. E. (1968). Laser display technology, *IEEE Spectrum* **5**(12), 39–50.

Baker, C. E. (1970). U.S. Patent No. 3,549,800.

Baker, C. E., and Alsabrook, C. M. (1967). A large screen color television laser display, *SWIEEECO Rec., Dallas, Tex.* pp. 5-7-1-5-7-8.

Baker, C. E., and Rugari, A. D. (1965). A large-screen real time display technique, *Proc. Soc. Inform. Display, Nat. Symp., 6th*, New York, pp. 85–101.

Baker, C. E., and Rugari, A. D. (1966). A large screen real time display technique, *Inform. Display* **3**, 37–46.

Beasley, J. D. (1971). Electrooptic laser scanner for TV projection display, *Appl. Opt.* **10**(8), 1934–1936.

Beiser, L. (1972). Opto-mechanisms for wideband laser scanning, *Proc. Electro-Opt. Syst. Design Conf., New York, Sess. VII*, pp. 255–265.

Beiser, L. (1974). Laser scanning systems, *In* "Laser Applications" (M. Ross, ed.), Vol. 2, pp. 53–159, Academic Press, New York.

Beiser, L., Lavender, W., McMann, R. H., Jr., and Walker, R. (1971). Laser-beam recorder for color television film transfer, *J. SMPTE (Soc. Motion Pict. Telev. Eng.)* **80**, 699–703.

Bennett, W. R., Jr., Knutson, J. W., Jr., Mercer, G. N., and Detch, J. L. (1964). Super-radiance, excitation mechanisms, and quasi-CW oscillation in the visible Ar$^+$ laser, *Appl. Phys. Lett.* **4**(10), 180–182.

Berg, A. D., Cormier, R. J., and Courtney-Pratt, J. S. (1974). High-resolution graphics using HeCd laser to write on kalvar film, *J. SMPTE (Soc. Motion Pict. Telev. Eng.)* **83**(7), 588–599.

Berlincourt, D. A., Curran, D. R., and Jaffe, H. (1964). *In* "Physical Acoustics" (W. P. Mason, ed.), Vol. 1, Part A, pp. 169–270. Academic Press, New York.

Black, J. F., Summers, C. J., and Sherman, B. (1972). Scanned-laser microscope for photoluminescence studies, *Appl. Opt.* **11**(7), 1553–1562.

Born, M., and Wolf, E. (1970). "Principles of Optics," Fourth Ed. Pergamon, Oxford.

Bridges, W. B. (1964). Laser oscillation in singly ionized argon in the visible spectrum, *Appl. Phys. Lett.* **4**(7), 128–130.

Brosens, P. J. (1971). Fast retrace optical scanning, *Electro-Opt. Syst. Design* **3**(4), 21–24.

Brosens, P. J., and Grenda, E. P. (1974). Applications of galvanometers to laser scanning, *Proc. Annu. Tech. Meet., Soc. Photo-Opt. Instrum. Eng., 18th, San Diego, Calif.*

Cann, M. W. P. (1969). Light sources in the 0.15-20-μ spectral range, *Appl. Opt.* **8**(8), 1645–1661.

Chang, I. C., Lean, E. G. H., and Powell, C. G. (1970). Dynamics of feedback-controlled Nd:YAG laser and its application for digital scan laser, *IEEE J. Quantum Electron.* **QE-6**(7), 436–441.

Chang, W. S. C. (1968). "Principles of Quantum Electronics," Ch. 8. Addison-Wesley, Reading, Massachusetts.

Chen, F.-S. (1970). Modulators for optical communications, *Proc. IEEE* **58**(10), 1440–1457.

Cheng, D., and Miller, R. C. (1973). A high-resolution laser micro-recording system with real-time viewing, *Dig. Tech. Pap., IEEE/OSA Conf. Laser Eng. Appl., Washington, D.C.* Pap. No. 10.6, pp. 58–59.

Chesler, R. B., Karr, M. A., and Geusic, J. E. (1970). An experimental and theoretical study of high repetition rate Q-switched Nd:YAlG lasers, *Proc. IEEE* **58**(12), 1899–1914.

Cohen, M. G., and Gordon, E. I. (1965). Acoustic beam probing using optical techniques, *Bell Syst. Tech. J.* **44**, 693–721.

Cohen, R. W., and Gorog, I. (1973). Frequency response of laser scanners and its optimization through apodization, *J. Opt. Soc. Amer.* **63**(9), 1071–1079.

Convert, G., Armand, M., and Martinot-Lagarde, P. (1964). Transitions lasers visibles dans L'argon ionisé, *C. R. Acad. Sci.* **258**, 4467–4469.

Coquin, G. A., Griffin, J. P., and Anderson, L. K. (1970). Wide-band acousto-optic deflectors using acoustic beam steering, *IEEE Trans. Sonics Ultrason.* **SU-17**(1), 34–40.

Crocetti, C. P., Calucci, E. J., Rugari, A. D., and Blank, C. M. (1968). Light valves, lasers and electroluminescent devices, *In* "Display Systems Engineering," Chap. 10, (Luxenberg H. R. and R. L. Kuen, eds.), pp. 319–391. McGraw-Hill, New York.

Culshaw, W., Kannelaud, J., and Peterson, J. E. (1974). Efficient frequency-doubled single-frequency Nd:YAG laser, *IEEE J. Quantum Electron.* **QE-10**(2), 253–263.

Dakss, M. L., and Powell, C. G. (1968). A fast digital scan laser, *IEEE J. Quantum Electron.*, **QE-4**(10), 648–654.

Davidovits, P., and Egger, M. D. (1971). Scanning laser microscope for biological investigations, *Appl. Opt.* **10**(7), 1615–1619.

DeMars, G., Seiden, M., and Horrigan, F. A. (1968). Optical degradation of high-power ionized argon gas lasers, *IEEE J. Quantum Electron.* **QE-4**(10), 631–637.

Dickson, L. D. (1972). Optical considerations for an acousto optic deflector, *Appl. Opt.* **11**(10), 2196–2202.

DiStefano, T. H., and Viggiano, J. M. (1974). Interface imaging by scanning internal photoemission, *IBM J. Res. Develop.* **18**(2), 94–99.

Dixon, R. W. (1967). Acoustic diffraceion of light in anisotropic media, *IEEE J. Quantum Electron.* **QE-3**(2), 85–93.

Dobbins, L. W. (1972). LR71 laser image scanner /recorder, *Proc. Ann. Tech. Meet. Soc. Photo-Opt. Instrum. Eng., 16th, San Mateo, California, 1972,* pp. 225–229.

Easton, R. A., Markin, J., and Sobel, A. (1967). Subjective brightness of a very short-persistence television display compared to one with standard persistence, *J. Opt. Soc. Amer.* **57**, 957–962.

Fowler, V. J. (1968). Laser color television projection systems, *In "Applications of Lasers to Photography and Information Handling"* (R. D. Murray, ed.) pp. 253–269. Soc. Photogr. Sci. Eng.

Fowler, V. J. (1974). Laser scanning techniques, *Proc. Ann. Tech. Meet., Soc. Photo-Opt. Instrum. Eng., 18th, San Diego, Calif.*

Fowler, V. J., and Schlafer, J. (1966). A survey of laser beam deflection techniques, *Appl. Opt.* **5**(10), 1675–1682.

Fowler, V. J., Stone, S. M., and Schlafer, J. (1971). Scanned laser techniques in color display and copying, 1971 *IEEE Intern. Conv. Dig.*, New York, pp. 172–173.

Gerig, J. S., and Montague, H. (1964). A simple optical filter for chirp radar, *Proc. IEEE* **52**(12), 1753.

Geusic, J. E., Bridges, W. B., and Pankove, J. I. (1970). Coherent optical sources for communications, *Proc. IEEE* **58**(10), 1419–1439.

Gordon, E. I. (1966). A review of acousto-optical deflection and modulation devices, *Proc. IEEE* **54**(10), 1391–1401.

Gordon, E. I., Labuda, E. F., and Bridges, W. B. (1964). Continuous visible laser action in singly ionized argon, Krypton, and Xenon, *Appl. Phys. Lett.* **4**(10), 178–180.

Gorog, I., Knox, J. D., and Goedertier, P. V. (1972a). A television-rate laser scanner-I. General considerations, *RCA Rev.* **33**(4), 623–666.

Gorog, I., Knox, J. D., Goedertier, P. V., and Shidlovsky, I. (1972b). A television rate laser scanner-II. Recent developments, *RCA Rev.* **33**(4), 667–673.

Gramenopoulos, N., and Hartfield, E. D. (1972). Advanced laser image recorder, *Appl. Opt.* **11**(12), 2778–2782.

Grenda, E. P., and Brosens, P. J. (1974). Closing the loop on galvo-scanners, *Electro-Opt. Syst. Design* **6**(4), 32–34.

Hance, H. V. (1964). Light diffraction by ultrasonic waves as a multiple-scattering process, *J. Acoust. Soc. Amer.* **36**, 1034–1035.

Hance, H. V., and Parks, J. K. (1965). Wide-band modulation of a laser beam using Bragg-angle diffraction by amplitude-modulated ultrasonic waves, *J. Acoust. Soc. Amer.* **38**, 14–23.

Harvey, A. F. (1970). "Coherent Light," Ch. 4. Wiley (InterScience), New York.

Hrbek, G., Lekavich, J., and Watson, W. (1970). An improved laser color TV system using acousto-optic interaction, *Dig. Pap., IDEA (Inform. Display, Evol. Advanc.) Symp., Soc. Inform. Display, New York*, pp. 40–41.

Hrbek, G., Lekavich, J., and Watson, W. (1971). An improved laser color TV system using acousto-optic interaction, *Proc. Soc. Inform. Display* **12**(2), 77–85.

Kahn, F. J. (1973). IR laser addressed thermo-optic smectic liquid crystal storage displays, *Appl. Phys. Lett.* **22**(3), 111–113.

Kenville, R. F. (1971). Noise in laser recording, *IEEE Spectrum* **8**(3), 50–57.

Kitaeva, V. F., Odintsov, A. N., and Sobolev, N. N. (1970). Continuously operating argon ion lasers, *Sov. Phys.—Usp.* **97**(3/4), 699–730.

Kitaeva, V. F., Osipov, Yu. I., Sobolev, N. N., Shelekhov, A. L., and Agheev, V. P. (1974). Probe measurements of Ar^+-laser plasma parameters, *IEEE J. Quantum Electron.* **QE-10**(10), 803–809.

Klein, W. R., and Cook, B. D. (1967). Unified approach to ultrasonic light deflection, *IEEE Trans. Sonics Ultrason.* **SU-14**(3), 123–134.

Korpel, A., Adler, R., Desmares, P., and Watson, W. (1966). A television display using acoustic deflection and modulation of coherent light, *Appl. Opt.* **5**(10), 1667–1675.

Korpel, A., Whitman, R. L., and Odom, M. (1971). Medium resolution scanning with an M-40R acousto-optic modulator, *Zenith Prod. Appl. Note* No. 1.

Lean, E. G. M., Quate, C. F., and Shaw, H. J. (1967). Continuous deflection of laser beams, *Appl. Phys. Lett.* 10(2), 48–50.

Lee, H. W. (1938). The Scophony television receiver, *Nature (London)* Vol. 142, No. 3584, 59–62.

Lee, H. W. (1939). Some factors involved in the optical design of a modern television receiver using moving scanners, *Proc. IRE* 27, 496–500.

Lee, T. C., and Zook, J. D. (1968). Light beam deflection with electrooptic prisms, *IEEE J. Quantum Electron.* QE-4(7), 442–454.

Kotovshchikov, G. S., Kuklev, V. P., Lantsov, N. P., Meerovich, G. A., Negodov, A. G., and Ulasyuk, V. N. (1974). Sealed scanned electron-beam-excited semiconductor laser, *Soviet J. Quant. Electron.* 4, 242–243.

Liberman, I., Larson, D. A., and Church, C. H. (1969). Efficient Nd:YAG CW laser using alkali additive lamps, *IEEE J. Quantum Electron.* QE-5, 238–241.

McMahon, R. E. (1971). Laser tests IC's with light touch, *Electronics* 44, 92–95.

Marantz, H., Rudko, R. I., and Tang, C. L. (1969). The singly ionized krypton ion laser, *IEEE J. Quantum Electron.* QE-5(1), 38–44.

Maydan, D. (1973). Laser-addressed light valves using liquid crystals, *Dig. Tech. Pap., IEEE/OSA Conf. Laser Eng. Appl., Washington, D.C.*, Pap. No. 18.6, pp. 89–90.

Melchior, H., Kahn, F. J., Maydan, D., and Fraser, D. B. (1972). Thermally addressed electrically erased high-resolution liquid crystal, *Appl. Phys. Lett.* 21(8), 392–394.

Montuori, J. S., Carnes, W. R., and Shim, I. H. (1973). Video-to-film color-image recorder, *Photogrammet. Eng.* 39(4), 395–400.

Murray, J. E., Pressley, R. J., Boyden, J. H., and Webb, R. B. (1974). CW mode-locked source at 0.532 μ, *IEEE J. Quantum Electron.* QE-10(2), 263–267.

NASA (1973). Laser scanner for testing semiconductor chips, *NASA Tech. Brief (Marshal Flight Cent.)* B73-10327.

Nowicki, T. (1974). A-O and E-O Modulators, Basics and Comparisons, *Electro-Opt. Syst. Design* 6(2), 23–28.

Okolicsanyi, F. (1937). The wave-slot, an optical television system, *Wireless Eng.*, pp. 527–536.

Packard, J. R., Tait, W. C., and Dierssen, G. H. (1971). Two-dimensionally scannable electron-beam-pumped laser, *Appl. Phys. Lett.* 19(9), 338–340.

Peters, C. J. (1965). Gigacycle-bandwidth coherent-light traveling-wave modulator, *Proc. IEEE* 53, 455–460.

Pinnow, D. A., Van Uitert, L. G., and Feldman, M. (1971). Photoluminescent conversion of laser light for black and white and multicolor displays-2: Systems, *Appl. Opt.* 10(1), 154–158.

Raamot, J., and Zaleckas, V. J. (1974). Laser pattern generation using X-Y beam deflection, *Appl. Opt.* 13(5), 1179–1183.

Randolph, J., and Morrison, J. (1971). Rayleigh-equivalent resolution of acoustooptic deflection cells, *Appl. Opt.* 10(6), 1453–1454.

Roberts, H. N., Watkins, J. W., and Johnson, R. H. (1974). High speed holographic digital recorder, *Appl. Opt.* 13(4), 841–856.

Robinson, D. M. (1939). The supersonic light control and its application to television with special reference to the scophony television receiver, *Proc. IRE* 27, 483–486.

Sawatari, T. (1973). Optical heterodyne scanning microscope, *Appl. Opt.* 12(11), 2768.

Schlafer, J., and Stone, S. M. (1969). Direct recording of color video images on film using scanned laser beams, *Symp. Image Display and Recording (SIDAR), Wright-Patterson AFB, Dayton, Ohio.*

Schreiber, W. F. (1973). Laser dry silver facsimile system, *Proc. Ann. TAPPI Reprogr. Conf., 3rd, Boston, Mass.* pp. 127–129.

Sieger, J. (1939). The design and development of television receivers using the scophony optical scanning system, *Proc. IRE* **27**, 487–492.

Silfvast, W. T. (1968). Efficient CW laser oscillation at 4416 Å in Cd(II). *Appl. Phys. Lett.* **13**(5), 169–171.

Silfvast, W. T. (1970). CW laser action on 24 wavelengths in Se II, *Appl. Phys. Lett.* **17**(9), 400–403.

Sliker, T. R., and Burlage, S. R. (1963). Some dielectric and optical properties of KD_2PO_4, *J. Appl. Phys.* **34**(7), 1837–1840.

Sommargren, G. E., and Thompson, B. J. (1973). Linear phase microscopy, *Appl. Opt.* **12**(9), 2130–2138.

Stone, S. M. (1967). Experimental multicolor real time laser display system, *Proc. Nat. Symp. 8th, Soc. Inform. Display, San Francisco, 1967*, pp. 161–168.

Stone, S. M., Schlafer, J., and Fowler, V. J. (1969). An experimental laser color TV projection display system, *Inform. Display* **6**(1), 41–44.

Taneda, T., Sato, T., Tatuoka, S., Aiko, M., and Masuko, H. (1973a). High-quality laser color television display, *J. SMPTE (Soc. Motion Pict. Telev. Eng.)* **82**, 470–474.

Taneda, T., Tatuoka, S., Aiko, M., Masuko, H., Yamamoto, M., Hashimoto, A., and Horiuchi, H. (1973b). A 1125-scanning-line laser color-TV display, *Dig. Tech. Pap., Soc. Inform. Display Int. Symp., New York* pp. 86–87.

Thouret, W. E., Strauss, H. S., Cortorillo, S. F., and Kee, H. (1964). High-brightness xenon lamps with liquid-cooled electrodes using standard lamp manufacturing techniques, *Illum. Eng. (New York)* **59**(9), 589–591.

Uchida, N., and Niizeki, N. (1973). Acousto-optic deflection materials and techniques *Proc. IEEE* **61**(8), 1073–1092.

Uchida, N., and Ohmachi, Y. (1970). Acousto-optical light deflector using TeO_2 single crystal, *Jap. J. Appl. Phys.* **9**(1), 155–156.

Ueda, M., Sato, T., Nawa, O. A., and Seo, T. (1968). On the two dimensional light modulator and its applications, *Trans. Soc. Instrum. Contr. Eng.* **4**(3), 280–286. (In Japanese).

Van Uitert, L. G., Pinnow, D. A., and Williams, J. C. (1971). Photoluminescent conversion of laser light for black and white and multicolor displays-1: Materials, *Appl. Opt.* **10**(1), 150–153.

Waksberg, A., and Wood, J. (1972). An automatic optical bias control for laser modulators, *Rev. Sci. Instrum.* **43**(9), 1271–1273.

Warner, A. W., White, D. L., and Bonner, W. A. (1972). Acoustooptic light deflectors using optical activity in paratellurite, *J. Appl. Phys.* **43**(11), 4489–4495.

Watson, W. H., and Korpel, A. (1969). Equalization of acoustooptic deflection cells in a laser color TV display, *Appl. Opt.* **9**(5), 1176–1179.

White, A. D., and Rigden, J. D. (1962). Continuous gas maser operation in the visible, *Proc. IRE (Corresp.)* **50**, 1697.

Wikkenhauer, G. (1939). Synchronization of Scophony television receivers, *Proc. IRE* **27**, 492–496.

Williams, R., and Woods, M. H. (1972). Laser-scanning photo-emission measurements of the silicone-silicone dioxide interface, *J. Appl. Phys.* **43**(10), 4142.

Woywood, D. J. (1972). Laser recording system evaluation, *Opt. Acta* **19**(12), 973–982.

Yamada, Y., Yamamoto, M., and Nomura, S. (1970). Large screen laser color TV projector, *Proc. Int. Quantum Electron. Conf., 6th, Kyoto*, pp. 242–243.

Yamamoto, M. (1974). U.S. Patent No. 3,818,129.

Yamamoto, M. (1975a). A 1125-scanning-line laser color TV display, *Hitachi Rev.* **24**, 89–94.

Yariv, A. (1967). "Quantum Electronics," Ch. 14. Wiley, New York.

Yamamoto, M., Tomiyama, S., Hashimoto, A., and Saito, S. (1971). U.S. Patent No. 3,617,933.

Yellin, M. (1972). Scanners and recorders for imagery transmission, *Proc. Annu. Tech. Symp. Soc. Photo-Opt. Instrum. Eng., 16th, San Mateo, California, 1972*, pp. 259–265.

Zook, J. D. (1974). Light beam deflector performance: A comparative analysis, *Appl. Opt.* **13**(4), 875–887.

Zworykin, V. K., and Morton, G. A. (1954). "Television," 2nd Ed., Ch. 5. Wiley, New York.

Display Applications of PLZT Ceramics

J. R. Maldonado, D. B. Fraser, and A. H. Meitzler*

BELL LABORATORIES
MURRAY HILL, NEW JERSEY

* Present address: Scientific Research Staff, Ford Motor Company, Dearborn, Michigan.

I. Introduction

A. Historical Background of PLZT Ceramics

Lead zirconate–titanate ceramic materials have been extensively investigated over the past 20 years, primarily in connection with their use as materials for electromechanical applications. Divalent and trivalent cations of many metallic elements have been added in small amounts to enhance both mechanical and piezoelectric properties of one sort or another and a large number of these doped lead zirconate–titanate systems have, been studied in detail.

Doped lead zirconate–titanate ceramics were investigated from the point of view of their usefulness as piezoelectric transducers or electromechanical filters. The first report of the potential usefulness of these materials in electrooptic devices was presented by Land (1967) at the IEEE International Electron Devices Meeting. Land reported that thin sections of bismuth-doped, lead zirconate–titanate materials could be made with low optical loss, and that the scattering of light beams passing through thin sections of these materials could be markedly altered by changing their state of remanent polarization. Since the state of remanent polarization could readily be switched in localized regions in thin, plate-shaped samples by the proper arrangement of electrodes, the possibility of realizing optical memory devices, electrically addressed and optically read, was thereupon recognized.

This recognition of the potential usefulness of ceramic materials in electrooptic devices provided the impetus for the systematic investigation of lead zirconate–titanate systems with a variety of metallic dopants (Land and Thacher, 1969; Thacher and Land, 1969). Lanthanum, which had been used in the past as a dopant of lead zirconate–titanate materials for certain piezoelectric applications, was found by Haertling and Land (1971) to be an additive particularly effective in achieving a ceramic body with good optical homogeneity, low optical insertion loss, and a wide range of useful electrooptic characteristics (Haertling, 1971). For convenience, the lead lanthanum zirconate–titanate ceramics were given the designation PLZT ceramics.

Haertling and Land (1971) mapped out the electrooptic properties of the PLZT ceramics over most of the compositional range between the extremes of lead zirconate and lead titanate, and with La added up to 25 at.%. The electrooptic characteristics undergo large changes over the whole compositional range, and these are reviewed in a recent work by the Sandia group (Land *et al.*, 1974). At present, the most useful materials for electrooptic applications appear to be those with a 65/35 zirconium–titanium ratio. The most convenient way to discuss the properties of material of this compositional ratio as a function of La content is to use the conventional designation PLZT $X/65/35$ with "X" equal to the atom percent of La.

A brief discussion of the electrooptic properties of $X/65/35$ ceramics as a function of X will be given in Section D. It will be helpful to the reader to review first the basic electrooptic effects in PLZT ceramics, and then to examine briefly the basic structures used for image storage and display devices.

B.   Basic Electrooptic Effects in PLZT Ceramics

There are two electrooptic effects useful in realizing display devices. These two effects are illustrated in Fig. 1. In each case, the effect is illustrated for a laser beam passing through a thin plate of ceramic. The first effect involves controllable light scattering and the second effect involves controllable birefringence. In the case of the light-scattering effect, shown in Fig. 1(I), if a small detector aperture is used, less light will pass through the aperture when the polarization vectors of the ferroelectric domains lie in the plane of the plate than when the polarization vectors are parallel to the light path. (Although this statement is generally true, the usefulness of the high lanthanum composition PLZT ceramic in the scattering mode depends to a large extent on index of refraction changes associated with domain growth accompanying phase changes of the material, rather than simply changes in the orientation of the polarization vectors within domains. This point will receive a detailed discussion in Section III.) The change in the direction of polarization vectors is accomplished by modifying the manner in which the electrodes of Fig. 1(II) are connected to the voltage source.

In the case of the birefringence effect, the ceramic acts like a crystal with uniaxial birefringence, having a magnitude of birefringence dependent upon the magnitude of the remanent polarization. When the remanent polarization vector is perpendicular to the light path, the material has maximum birefringence; when it is parallel to the light path, the material has minimum (ideally zero) birefringence.

I. VARIABLE LIGHT
SCATTERER

II. VARIABLE BIREFRINGENT
ELEMENT

(a) TRANSVERSE MODE

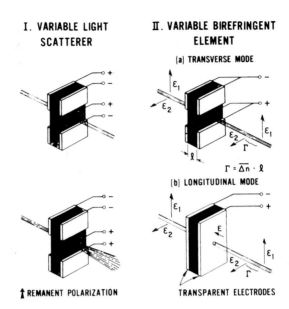

$\Gamma = \overline{\Delta n} \cdot \ell$

(b) LONGITUDINAL MODE

↑ REMANENT POLARIZATION   TRANSPARENT ELECTRODES

Fig. 1.   Electrooptic effects useful in realizing display devices. I. Variable light
scattering: the voltage is applied first between the two positive electrodes (joined together) and the two negative electrodes (joined together) in the upper figure. This
produces minimum light scattering. When the voltage is applied, as shown in the lower
figure, maximum light scattering is obtained. II. Variable birefringence. (a) Transverse
mode: The applied electric field is always in the plane of the figure. The direction of the
electric fields of the incident light beam are also shown. (b) Longitudinal mode: The
applied electric field E is always parallel to the direction of propagation of the light
(through the plate thickness). The directions of the electric fields of the incident light
beam are also shown.

An average value of birefringence, $\overline{\Delta n}$, is easily determined experimentally for the ceramic from measurable parameters by means of the
equation

$$\overline{\Delta n} = \Gamma/t \tag{1}$$

where Γ is the phase retardation measured for the incident light with
components $\mathcal{E}_1$ and $\mathcal{E}_2$ and t is the thickness of the plate. In plates of PLZT
6/65/35, values of $\overline{\Delta n}$ as large as 15×10^{-3} have been observed, making
it readily possible to achieve a change of a half wavelength of phase retardation in switching maximum remanent polarization to some intermediate state, with minimum $\overline{\Delta n}$ in plates as thin as 75 μm for 633-nm
light from a He–Ne laser.

In the case of the birefringence mode of operation, two different structures are useful. The most basic and obvious is that of the transverse,

light-gate structure shown in (IIa) of Fig. 1. In addition, it is possible to use the ceramic in a longitudinal mode of operation. Ordinarily, in birefringence devices using single crystals, the ability to operate in a longitudinal mode (applied field parallel to light propagation direction) depends upon the material anisotropy in the plane perpendicular to the direction of the applied field. A poled ceramic does not ordinarily exhibit this propperty because of the randomness of the domain orientations in the plane of the plate; but the stress sensitivity of PLZT materials makes it possible to use the technique of strain biasing (to be described in detail in Section II) which serves to effectively render the ceramic anisotropic. In this case again, plates as thin as 75 μm can provide half-wavelength changes of phase retardation for 633-nm light, with applied voltages as low as 100 V (Maldonado and Meitzler, 1971).

C. Changes with Time of the Assessment of the Relative Usefulness of the Two Basic Effects

As already mentioned, the earliest work done on lead zirconate–titanate materials for electrooptic device applications was done from the point of view of employing the variable light-scattering effect. The contrast ratio achievable with the early Bi-doped ceramic using the birefringence mode was poor, because incident polarized light was largely depolarized in passing through this material. When the PLZT materials were introduced, the situation changed. The depolarization accompanying scattering was reduced by making fine-grain ceramics with grain sizes of ∼2 μm, and it became possible to obtain higher contrast ratios (or ON/OFF ratios) using the birefringence effect than when using the scattering effect. Thus, for the period from about 1969 to 1971, the principal emphasis of PLZT device development was on devices using the birefringence effect. In connection with the development of image storage and display devices, Smith and Land (1972) recognized that under certain conditions a high-contrast image device could be achieved using the scattering mode of operation. The present state of the art is one in which one or the other mode of operation can be advantageous depending upon the details of the application.

D. Basic Principles of Ferroelectric-Photoconductor Image Storage and Display Devices

Even before the advent of PLZT materials, consideration had been given to the idea of combining a photoconductive layer and a thin plate of a ferroelectric material in layered-shaped structures, to take advantage of the localized switching and storage capabilities of ferroelectrics. For example, Hanlet (1963) disclosed a layered ferroelectric-photoconductor structure that is basic to many of the image storage devices using these materials. Interestingly enough, the objective of Hanlet's invention was not a display device, but rather a solid-state electronic camera.

The combination of a thin plate of PLZT ceramic and a photoconductive film, along with the requisite transparent electrodes, was first employed in image storage devices using the birefringence effect; however, the same basic arrangement is equally applicable to image storage devices using the scattering mode. The two basic modes of operation differ principally in the details of the image viewing systems and not the image formation systems. A simple form of ferroelectric-photoconductor device, called a ferpic,* is shown in Fig. 2.

1. *The Rudimentary Ferpic*

The rudimentary ferpic uses a 50-μm-thick piece of lead zirconate–lead titanate ceramic† having a grain size of approximately 1 μm. Originally, the plate is poled to have a remanent polarization in the plane of the plate (L-state) by means of an applied field of ∼20 kV/cm. The sample is poled before the photoconductive film and the transparent electrodes are applied. The specific poling conditions used for a given plate are adjusted to give a state of remanent polarization causing a half wavelength of phase retardation when polarized light is transmitted at normal incidence through the plate.

After the plate is initially in a condition such that all regions of the plate in the area used for storing the picture are in the L-state, the photoconductive film and transparent conductive films are applied. In the version of a ferpic outlined in Fig. 2, the photoconductive film (PVK)‡ is applied simultaneously to both sides of the plate by a dip-coating technique. Transparent conductive electrodes are next applied. (In the original experimental devices, half-transparent films of Cr–Au were vapor-deposited on the two surfaces; in actual devices, more transparent electrodes of tin-oxide or indium oxide would be preferred.) Fine wire leads are attached to

---

* Image storage and display devices using the birefringence effect in PLZT were first realized at Bell Laboratories and named "ferpics" by the group working on them (Meitzler *et al.*, 1970). Image storage and display devices using the scattering effect were first realized at Sandia Laboratories and named "cerampics" by their developers (Smith and Land, 1972a).

† The ceramic used in our original experiments was produced by the Clevite Corporation, Cleveland, Ohio.

‡ The abbreviation PVK stands for polyvinyl carbazole which will be described in Section VI. The material used was obtained from Polyscience, Inc. P.O. Box 4, Rydal Pennsylvania.

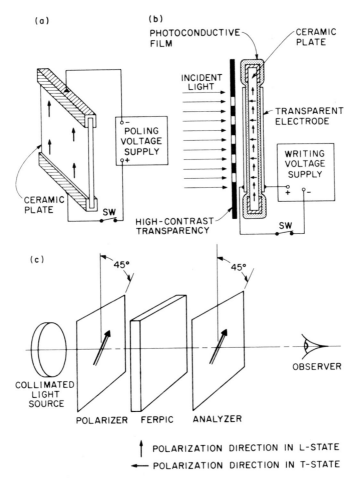

Fig. 2. (a) Prepoling of rudimentary device. (b) Arrangement used to write the picture information into the ceramic plate. (c) Arrangement used to make the stored image visible by placing the device between crossed polarizers.

these electrodes and are used to connect the device to the voltage source used to supply a switching field in the thickness direction.

Figure 2, in addition to showing the structural features of the device, shows the arrangement used to write the picture information into the ceramic plate. A high contrast transparency is placed immediately in front of the ferpic and is illuminated with collimated light. The voltage supply is pulsed on. In the regions where the light passes through the transparency, the photoconductive film becomes conductive and the field in the ceramic becomes large enough to produce switching in the form of 90° average polarization rotation (see Maldonado and Meitzler (1972) for allowed switching angles in rhombohedral materials). This mode of operation is described by Land and Thacher (1969) in connection with light-gate structures using various configurations of metal electrodes. In ferpic devices, the 90° average polarization-rotation mode of operation is obtained by the use of photoconductive films and transparent conductive films working in combination with metal electrodes.) Localization of the switching field requires that the dielectric constant of the ceramic, K_{PZT}, be much larger than the dielectric constant of the photoconductor, K_{PVK}. For the materials used in the early experimental devices, $K_{PZT}/K_{PVK} \approx$ 400. For a 50-μm-thick ceramic plate, the writing conditions were the following: (1) a white light flux of 2 mW/cm², (2) a 200-V supply, and (3) a voltage pulse duration of $\sim$1 min.

As already described, the stored image can be made visible by inserting the ferpic between a polarizer and an analyzer, as indicated in Fig. 2c, and illuminating it with light from a collimated, monochromatic light source. (The degree of collimation and monochromaticity involved are not critical. Most of our experimental work was done using white light from ordinary incandescent sources and, in fact, the two photographs of Figs. 3 and 7 were made with this sort of light.) If the phase retardation produced by regions in the L-state is $\lambda/2$, efficient use of light is obtained when polarizer and analyzer are set parallel to each other and at an angle of 45° to the electric polarization vector in the ceramic. The regions in the L-state appear opaque; the regions switched to the T-state present no birefringence to the incident light and appear as bright areas to the viewer.

The ferpic just described can be used as a kind of photographic plate capable of two-level (black and monochrome) image storage with a respectable degree of resolution. Figure 3 shows photographs made using a low-power microscope to view a pattern from a resolution test chart stored in a 50-μm-thick ceramic plate. The number of lines per millimeter in the individual columns marked A, B, C, D, and E is indicated in the caption.

This photograph demonstrates that resolutions better than 30 lines/mm are obtainable under the described conditions of operation. A number of experiments have been carried out to establish the principal factor limiting the resolution. Fringing fields within the 50-μm-thick plate appear to be the principal factor, since both the optical techniques and the photoconductor used have been demonstrated to have at least an order of magnitude finer resolution capabilities. The observed resolution is already equivalent to 2–3 cycles of variation over a distance equal to the thickness of the plate.

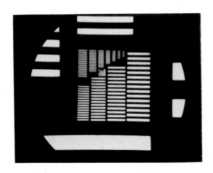

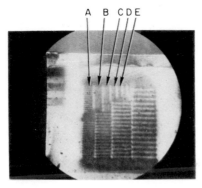

Fig. 3. Photograph made using a low-power microscope to view a pattern from a resolution test chart (shown on the top of the figure) stored in a 50-μm-thick ceramic plate. The number of lines per millimeter in the individual columns marked A, B, C, D and E is as follows: A, 25.1; B, 31.6; C, 39.8; D, 50.1; E, 63.1.

The device described in Fig. 2 is not a practical display device since, like a photographic plate, it can be used only once. Breakdown in the photoconductive film through the transparent electrodes on the surface prevent the polarization vector under these electrodes from being switched back to the L-state when a voltage is applied to the poling electrodes. We will next consider a form of ferpic that offers the capability of being electrically changeable.

2. *The Interdigital-Array Ferpic*

An interdigital electrode array deposited on one side of a ceramic plate provides, in principle, a means of switching the polarization vectors back into the plane of the plate after they have once been switched normal to it. An exploded view of the layer structure proposed for use in a changeable ferpic is shown in Fig. 4. In addition to enabling the plate to be switched into the L-state, use of the interdigital array has the great practical advantage of reducing the voltage required to pole in the longitudinal direction. As already indicated, 20 kV is required to pole over a distance of 1 cm. If the elements of the array have the same spacing as the plate thickness, the same voltage supply ($\sim$200 V for a 50-μm-thick plate) can be used to establish both longitudinal and transverse switching fields. The obvious disadvantage to this approach is that the stored image will now be broken into a number of discrete lines. The extent to which the presence of the lines is evident and objectionable will depend on the details of use.

The ferpic structure shown in Fig. 4 constitutes an electrically changeable image storage and display device that functions in two steps: (1) RESET, and (2) WRITE. The first two steps are illustrated in Fig. 5; the third step involves essentially the same elements shown earlier in Fig. 2c. In the RESET step, the electrodes of the interdigital array are connected to the voltage supply, and the resulting field switches the remanent polar-

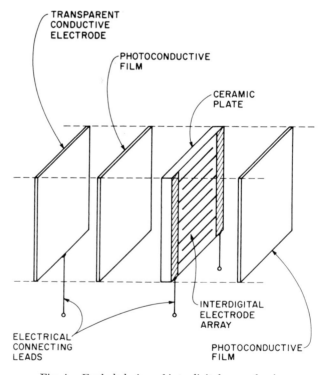

Fig. 4.   Exploded view of interdigital-array ferpic.

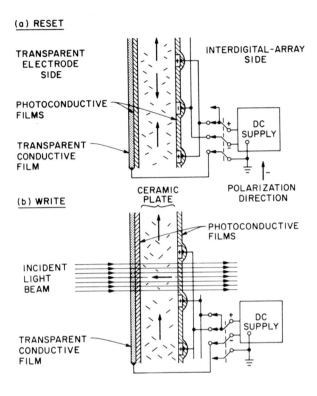

Fig. 5.   (a) Reset and (b) write steps.

ization vectors predominantly into the plane of the plate. In this condition of polarization (L-state), every region of the plate has maximum bire-fringence for linearly polarized light incident normally. In the WRITE step, the elements of the array are connected in parallel to one terminal of the supply and the other terminal is connected to the transparent conductive electrode. Light is directed at the area to be switched causing the photo-conductive layers to conduct and the electric field in the ceramic under the illuminated area to exceed the coercive field.

If the image to be stored in the ferpic is broken into elements, the spacing between lines of the interdigital array should be less than or equal to the size of an element. Writing the image, an element at a time, as in a tele-vision picture, can be accomplished by modulating either the addressing light beam or the power supply. It is equally conceivable that the picture could be formed all at one time by projecting some desired image on the plate and switching the illuminated elements.

The basic ideas of the interdigital-array device were first demonstrated in a device structure using PVK films. The details of an early interdigital-array ferpic are shown in Fig. 6, and an example of the image storage obtained is shown in Fig. 7. The upper figure shows the original, simple, high-contrast image. The lower figure shows the image observed through a low-power polarizing microscope. The image was stored in a 50-μm-thick ceramic plate with a square working area 0.8 cm on a side.

The experimental structure of Fig. 6 differed from the original structure described in Fig. 4 by the inclusion of an additional transparent conductive electrode on the array side of the ferpic. This additional film was needed because the PVK films did not have a high enough conductivity when illuminated to establish an equipotential region between the elements of the electrode array. While the addition of the transparent conductive electrode to the array side of the ferpic solved the problem of the low conductivity in the PVK, this modification had the disadvantage that it hindered erasure of the stored image. The PVK film between the transparent electrode and the array is now subject to breakdown field strengths when the RESET voltage is applied. In addition, experiments with PVK

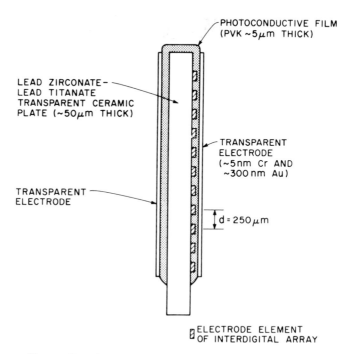

Fig. 6.   Details of interdigital-array ferpic using PVK films.

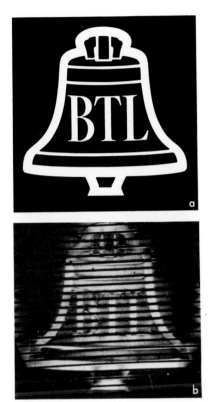

Fig. 7.   Photograph of an image stored in an early interdigital-array ferpic device.
(a) Photograph made from the original negative used to write the stored picture, (b).

films on devices using the 90° polarization rotation have indicated that the
PVK film constrains the motion of domains, probably through a mechanism of trapping polarization charges at the PZT–PVK interface.

3. *The Strain-Biased Ferpic*

As long as the technique used to store an image in a piece of ceramic
involved an operation equivalent to rotating domains about 90° by the
sequential application of two fields, at 90° to one another, ferpics had
serious disadvantages; e.g., the high field required to put the rudimentary
ferpic in the L-state or the grid of electrodes required to switch from T- to
L-shape in the interdigital-array ferpic. These disadvantages were cir-
cumvented by the invention of the technique of strain biasing (Maldonado
and Meitzler, 1971; Meitzler and Maldonado, 1971). The basic structure
of a strain-biased ferpic is shown in Fig. 8.

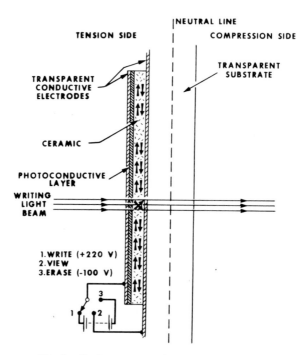

Fig. 8.   Basic structure of strain-biased ferpic.

PLZT materials are ferroelastic as well as ferroelectric; i.e., mechanical stresses are effective in achieving a high degree of domain orientation. Domains tend to align themselves along a tension axis. Thus, if a thin plate of PLZT is bonded to a ferroelectric substrate, and the substrate is flexed in such a way that the material is strained uniformly with the tension axis in the same direction over the active area, domains will switch their polarization parallel to the tension axis. Since there is no net applied field, the polarization vectors tend to align in antiparallel pairs in order to keep the average remanent polarization zero. This condition is shown in Fig. 8. Strain biasing renders the material birefringent, with polarization vectors in a condition analogous to the L-state of the previous two devices described in this section. Application of an electric field causes domains to switch under the region of the photoconductive layer illuminated by high-intensity, incident light, such as the laser beam shown in the figure. The material switched out of the initial condition is switched into a condition of lower birefringence analogous to the T-state. If a voltage of $+V$ is applied to the terminals to write and store an image, the stored image, once written, can be erased by the application of $-V/2$ to the terminals and simultaneously flooding the photoconductor with light. Practical devices operate with thin plates 50–75 μm thick and operate with voltages as low as 100 V.

The advantages of the strain-biasing technique are that it allows for devices (1) capable of operating at relatively low voltages, and (2) having large ($\sim$1–4 cm^2) storage areas free of interfering electrode lines. There are disadvantages associated with the necessity of forming thin, dirt-free bonding layers and straining the ceramic to relatively high states of tensile strain ($\sim$10^{-3}). Since this is the most attractive form of birefringent ferpic for image storage and display applications, the details of design and operation of this device structure will be discussed in Section II.

4. Scattering-Mode Image Storage and Display Devices

a. The Basic Cerampic. The simplest form of device structure—the one shown for the rudimentary ferpic—is capable of being used as a cerampic operating with fields applied only in the thickness direction. In the case of light scattering, the operation is possible without strain bias because only a random orientation of domains in the plane of the plate is needed in the L-state. This is in contrast to the birefringence case in which an optic axis in the plane of the plate is necessary for operation, and it is obtained by means of strain biasing. The cerampic structure is shown in Fig. 9a. The photoconductive layer is used in the same way as in the ferpic devices to modulate the intensity of the switching field in the ceramic during the operation of image storage. There are again two basic end states of the material: (1) an L-state in which the domains are now lying in the plane of the plate, but randomly oriented, and (2) a T-state in which the domains are aligned predominantly in the direction of the applied field.

In operation, the active area of the plate is first switched to the T-state by applying a voltage of $+V$ to the terminals. This aligns the domains in the direction of the applied field. To write the image into the device, a voltage of $-V/2$ is applied to the terminals; the image may be stored all at once, as in the use of contact exposure shown in Fig. 7, or written in element by element by a deflected laser beam. When the domain polarization vectors (as shown in Fig. 9b) are in condition A (the T-state), the material scatters light least. Material in condition B scatters light over large angles. The plate, when placed in an ordinary optical projection system, will form an image on a screen with regions A producing bright picture elements and regions B producing the dark picture elements. In many cases, an image of adequate contrast can be formed with the projection lens itself serving as the exit aperture of the projection system, limiting the angle over which the scattered light rays are collected.

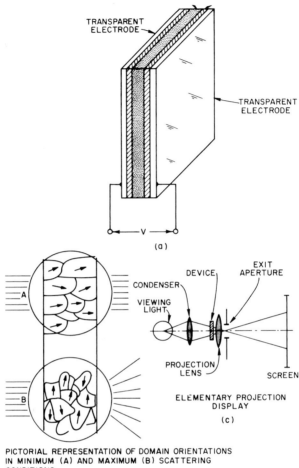

(a)

(b)

(c)

PICTORIAL REPRESENTATION OF DOMAIN ORIENTATIONS
IN MINIMUM (A) AND MAXIMUM (B) SCATTERING
CONDITIONS.

Fig. 9. (a) Scattering-mode ferroelectric-photoconductor device. (b) Pictorial representation of domain orientations in minimum A and maximum B scattering conditions. (c) Elementary projection display.

The cerampic has several outstanding advantages: (1) the useful form of the device has a simple structure that does not require an elaborate support structure (as in the case of the strain-biased ferpic), (2) no polarizer or analyzer is required to display the image, and (3) broadband (white) light can be used to project the image. (White light can only be used with a ferpic if elaborate compensation techniques are used.) A more detailed discussion of the structure and performance features of cerampics will be given in Section III.

b. The Fericon. In a paper by Land and Smith (1973) a new form of scattering-mode display device, called a Fericon, was announced. In this case, however, the device functions in reflection rather than transmission, and makes use of the surface deformations that accompany changes in polarization rather than changes in the index of refraction. The device has nearly the same form as the basic cerampic shown in Fig. 9a with the differences that the transparent electrode on one side is replaced by an opaque metallic film, and the transparent electrode is modified by the formation of a Ronchi grating on its surface. Typical performance figures reported by Land and Smith are contrast ratios of 5:1, resolution of 60 line pairs/mm and insertion loss of ∼3 dB. This new device structure has not as yet been investigated in detail; consequently, it is difficult to assess its eventual usefulness. As a general rule, however, reflective mode display devices require more complicated, less efficient optical projection systems and are, therefore, not as attractive for commercial or consumer applications.

E. LANTHANUM AND TEMPERATURE DEPENDENCE OF ELECTROOPTIC EFFECTS IN $X/65/35$ PLZT CERAMICS

The PLZT family of ceramics and the related electrooptic devices provide an outstanding example of the beneficial interaction between the development of device concepts and the development of new materials. Initially, the recognition of the potential usefulness of lead zirconate–titanate ceramics stimulated work to improve the optical properties of these materials, thus leading to PLZT ceramics. As new, unexpected effects were discovered in the PLZT ceramics, ideas for new types of devices were stimulated. In all likelihood, the final turn of this circle of events still has not taken place.

The earliest attempts at realizing electrooptic devices with intrinsic memory capabilities involved working with the material in a compositional region where the material is in the ferroelectric, rhombohedral phase. Rhombohedral-phase, lead zirconate–titanate materials combine relatively large values of polarization (remanent polarizations as large as 35 $\mu C/cm^2$), with relative ease of switching (coercive fields less than 10 kV/cm), and good lifetime. Early in the work with PLZT ceramics, the Zr/Ti ratio of 65/35 was determined to be a favorable region and almost all of subsequent display device work was done with material having the compositional designation $X/65/35$ (La/Zr/Ti).

The earliest device work was done with relatively low doping levels of La; e.g., $X = 2$. Many of the papers published in the period 1969–1970 deal with this material. However, the addition of larger concentrations of La to the ceramic made it possible to obtain good optical properties along with enhanced sensitivity to applied electrical or mechanical fields (Haertling and Land, 1972). At first, this favorable combination of circumstances was associated principally with the fact that the material was close to a morphotropic phase boundary. More recent work by Meitzler and O'Bryan (1973) led to a recognition that the high ferroelastic sensitivity of the material leads to the creation of a polymorphic phase which allows the material to achieve a high level of induced polarization with large electro-optic (birefringence and scattering) effects. A closely related phenomenon, called penferroelectricity, allows the material to be used for light-valve display devices using the transverse electrooptic effect. In this use, the material is almost ferroelectric at room temperature; it has a noncubic structure but a polarization that is so low that the material does not support

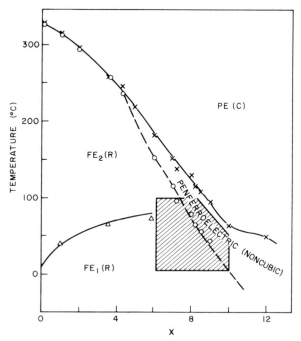

Fig. 10. Phase diagram for $X/65/35$ ceramics. $FE_1(R)$ and $FE_2(R)$ stand for the low temperature and high temperature rhombohedral phases, respectively. PE(C) stands for the paraelectric cubic phase. The shaded area enclosed by the irregular pentagon is the region of most interest for display applications.

domain walls. Under the influence of low-level fields, either mechanical or electrical, the polarization increases and this increase is accompanied by the generation of domains within the materials. In the absence of applied fields, the material is optically isotropic with low losses because the domain walls that cause scattering are absent. Applying a field causes domains to grow, and this effect enhances both the ferroelastic and electrooptic sensitivity of the material.

The changes in crystalline structure of the PLZT ceramics as functions of La content and temperature are fundamental to the usefulness of these materials. The phase diagram of Fig. 10 summarizes the structural dependence on temperature and La content for the $X/65/35$ ceramics. The symbols $FE_1(R)$ and $FE_2(R)$ stand for the low temperature and high temperature rhombohedral phases, respectively. $PE(C)$ stands for the paraelectric cubic phase. The region labeled penferroelectric (noncubic) is the region in which the material no longer supports domain walls, although it is highly sensitive to electrical or mechanical fields. The shaded region, enclosed by the irregular pentagon, is the region of most interest for display device applications, since material in this region combines low optical insertion loss and good optical homogeneity with high sensitivity to applied fields. As of 1973, most display applications of PLZT materials were concentrated on the materials and temperature ranges within this relatively small region. These applications will be described in the later sections of this chapter.

II. Birefringence-Mode, Strain-Biased, Image Storage and Display Devices—Ferpics

Two types of devices for image storage and display have been constructed utilizing the strain-biasing technique: one type operates in a transmission mode (Maldonado and Meitzler, 1971), and the other type operates in a reflection mode (Maldonado and Anderson, 1971). The transmission mode device will be discussed first.

A. Transmission Mode Devices

1. *Device Design Details*

The device structure for the transmission device is shown in Fig. 11. Three types of photoconductive films have been used in this kind of device: PVK films, CdS films, and a mixture of $Zn_xCd_{1-x}S$ sputtered film. All three of them will be described in detail in Section VI. The steps followed in the fabrication of the devices using the PVK films are briefly described below.

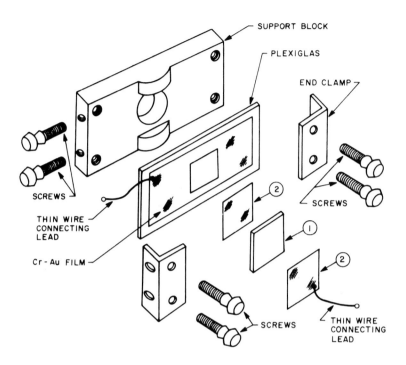

Fig. 11.   Device structure for strain-biased transmission mode device. 1. PLZT ferroelectric ceramic plate; 2. transparent In_2O_3 electrode.

A thin plate of ferroelectric ceramic is optically polished to a final thickness of ~75 μm. The plate is annealed above its Curie point, and a transparent electrode of indium tin oxide (Fraser and Cook, 1972) is deposited on one side. The ceramic plate with the transparent electrode film facing down is epoxy-bonded to a Plexiglas plate ~3 mm thick. A conductive Cr–Au film is deposited on the Plexiglas plate before bonding to make contact with the transparent electrode on the ceramic plate, at the outer edge of the plate. The outer side of the ceramic plate is coated with the PVK film. A transparent electrode (thin gold or indium tin oxide) is deposited on the PVK film. Thin wires are soldered to the Cr–Au film on the Plexiglas plate and to the top of the transparent electrode on the photoconductor.

In the devices with the $Zn_xCd_{1-x}S$ mixture, the photoconductive film and the top transparent electrode (indium tin oxide) are deposited on the ceramic plate prior to the bonding operation.

2. Dependence of Birefringence Changes On Strain Biasing

a. Strain Biasing the Ceramic Plates.   The ceramic plate of a strain-biased ferpic is uniformly strained by bending the Plexiglas substrate as shown in Fig. 12. If the ceramic plate is mounted on the outer surface of the Plexiglas plate, a tensile strain is obtained along a direction parallel to the long dimension of the Plexiglas substrate. (This statement is made referring either to Fig. 11 or Fig. 12.) If the ceramic plate is mounted on the inner surface of the Plexiglas plate, a compressive strain is obtained along the length direction. In either of the two mounting positions, experimental observations have verified that a uniform condition of birefringence is obtained in the ceramic plate when the Plexiglas plate is flexed.

Straining the plate as indicated produces a state of birefringence that results mainly from the realignment of domains in a direction parallel to the tension axis. In Fig. 13, the phase retardation—measured in units of

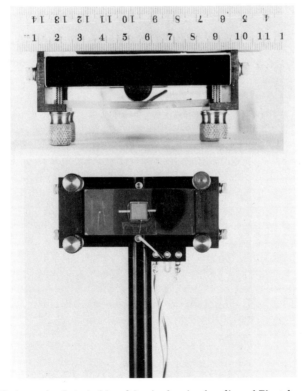

Fig. 12.   Photograph of strain-biased ferpic showing bending of Plexiglas substrate.

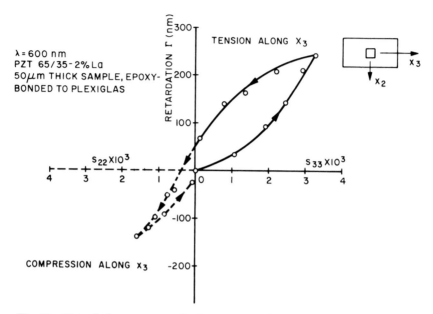

Fig. 13.   Retardation versus strain for a 75-μm-thick plate of PLZT 2/65/35.

fractions of a wavelength for 633-nm light normally incident on a 75-μm-thick plate of PLZT 2/65/35—is plotted as a function of strain measured with a bonded resistance strain gauge. The arrows indicate the path taken by the sample, starting from an initial state with almost no birefringence. The strain is increased to a maximum strain of close to 3×10^{-3}, and then decreased to zero. Note that upon returning to zero strain the sample has an appreciable remanent birefringence, indicating a permanent realignment of domains. Since this domain realignment or switching is produced without the application of an electric field, the resulting domain configuration is thought to be one in which the switched domains align parallel to the tension axis, but in pairs antiparallel to one another, in order that the macroscopic polarization vector in the plane of the plate remains zero (since Δn is an even function of the remanent polarization P_R, the fact that the average $\langle P_R \rangle = 0$ does not preclude nonzero Δn).

   b. *Birefringence Changes Produced by Applied Electric Fields.*   To study the birefringence changes accompanying domain switching in a strain-biased ferpic, we prepared a bonded plate of PLZT 2/65/35 without the photoconductor, and measured remanent birefringence (at zero applied voltage) as a function of switching voltage. These data are shown in Fig.

14. Two different modes of operating the same sample are illustrated by the data of this figure. In the first mode, the principal strain of the plate is tension, and in the second mode, the principal strain is compression. The vertical scale measures phase retardation (in nanometers) for the two respective cases, and the horizontal scale measures voltage applied to the transparent electrodes during switching. The voltages were applied for long enough times to allow the sample to switch under voltage-controlled conditions (Maldonado and Meitzler, 1970) before being removed for the birefringence measurements.

Discussion of the data of Fig. 14 is aided by the introduction of a reference coordinate system. This is seen in Fig. 15. Note that the x_1 axis coincides with the thickness direction of the ceramic plate. The x_3 axis coincides with the tension axis in the case when tension is the largest strain applied to the ceramic plate and coincides with the compression axis in the case when compression is the largest strain. The experimental arrangement results in the coordinate axes coinciding with the principal axes of strain in the ceramic plate, with S_{11}, S_{22}, and S_{33} being the corresponding principal components of strain. An analysis of the state of strain

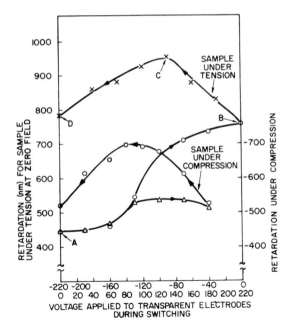

Fig. 14.   Retardation Γ versus switching voltage (zero field) for a PLZT 2/65/35 ceramic 75 μm thick. Data for the sample under tension and compression are shown in the figure.

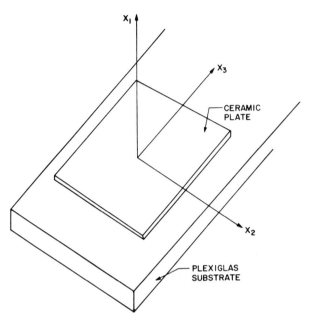

Fig. 15. Reference coordinate system. The x_3 axis coincides with the tension (compression) axis.

induced in the ceramic plate by the experimental conditions is presented in Maldonado and Meitzler (1971).

We will consider the case of tension first, i.e., where S_{33} is the largest strain component and has a positive sign. After the sample is strained and before any voltage is applied, the sample is in the condition indicated by point A. When voltage is applied and increased in steps, the measured retardation follows the path A to B. Surprisingly, the birefringence of the sample increases upon the initial application of electric field in the thickness direction. In the vicinity of B, the retardation levels off and an increase of the field in the direction originally applied causes no further increase; in fact, voltages larger than 220 V cause the birefringence of this particular sample to decrease. At point B, the applied field is about 30 kV/cm, which is slightly more than the 25 kV/cm required to produce saturated remanent polarization in a free plate. With the sample at B, the voltage is reversed and increased from zero in steps. (This change in the polarity of the applied voltage is the reason for the two voltage scales along the abscissa of the graph.) The initial increase of voltage in the reverse direction is accompanied by an additional increase in retardation. The retardation is observed to go through a maximum at C and return to a level of retardation at D, approximately equal to that of the sample at B. Repeated reversals of the polarity of the applied voltage enable the sample to be cycled back and forth between B and D; however, in use, the sample is cycled only between points B and C, since these two states give the maximum change in birefringence useful for image storage applications, for minimum applied voltages.

The behavior of the sample when subjected to a principal stress of compression along the x_3 axis is also shown in Fig. 14. The data for compression were obtained by taking a sample similar to the one used for tension and subjecting it to a compressive strain large enough to produce an initial phase retardation equal in magnitude but opposite in sign to that produced in the tension case. Note that the retardation of the sample saturates at a lower level in compression than in tension, but that the magnitude of the reversible retardation change is substantially the same for the two cases.

The fact that both the x_1 and x_2 axes are tension axes causes the domains to align predominantly in a plane normal to the x_3 axis, but again in such a manner that the average macroscopic polarization vector is zero. Application of 220 V to the electrodes results in a switching field along the x_1 axis and causes more domains to switch into a direction along the x_2 axis than switch out of it. The change in retardation with the initial application of voltage is less because now the applied field and a tension axis are both in the x_1 direction, making it easier for the domains to switch into this direction, and thereby decreasing the number of domains available to align along the x_2 axis. However, when -100 V is applied, the number of domains available to realign along the x_2 axis is the same as in the tension case; this fact is believed to account for the observation that the variation of phase retardation between maximum and minimum values is the same for compression as for tension.

The fact that the range of change in birefringence is the same for tension and compression suggests that the number of domains free to switch is the same in both cases. The results suggest that there is probably an optimum level of strain for a given thickness of a given material. Detailed studies of this have not been made; however, studies made in strain-biased transverse light gates have shown that for large strains a saturation occurs due to domain locking. This is because of the high internal fields developed by the piezoelectric strain. Roberts (1972) has also reported some saturation of the switched birefringence as the tensile strain is increased above 3×10^{-3}.

3. Writing and Viewing the Stored Image

Let us direct our attention back to the curve for the case of a tensile principal strain (as shown in Fig. 14) and think in terms of an actual device including a photoconductive film. In writing an image into the ceramic, the preferred mode of operation is to start with the ceramic in the L-state (C) and switch, under the control of light, selected regions into the T-state (B). A voltage pulse of ~100 V is required to switch the illuminated regions from the L-state to the T-state for the 8% La composition. The writing operation may be done by projecting an image onto the ceramic plate, by a swept-beam technique, or by using a simple contact-printing technique. For maximum contrast and minimum insertion loss, a half-wavelength change in retardation for the regions of the plate switched from C to B would be optimum. The data of Fig. 14 show only a quarter-wavelength change. This amount of change is enough to render visible a stored image; however, the optical insertion loss of the device is larger and the contrast ratio is less than would be obtained if the device had a full $\lambda/2$ change in phase retardation. One way of obtaining a full $\lambda/2$ change with the device of Fig. 14 is to use it in a reflection mode of operation.

4. *Erasing the Stored Image*

Let us again refer to Fig. 14 and direct our consideration to the curve for the tension case. A voltage of approximately -100 V needs to be applied across the electrodes in order to accomplish erasure, which amounts to returning the whole plate to the L-state (C). This is most simply accomplished by flooding the photoconductive film covering the whole ceramic plate with light during the time that the voltage is on. However, even though the strain-biased ferpic structure is only a two-terminal device, it is not necessary to erase the whole image at once. One of the advantages of the strain-biased ferpic is that it is ideally suited to working in a "periodic up-date" mode of operation. This is illustrated by Fig. 16, which shows how only a localized region addressed by a writing light beam can be written into or erased, depending on the position of the voltage switch. The earlier versions (Meitzler *et al.*, 1970) of ferpic devices did not have this feature. The best one could do with the interdigital array device would be to break up the array into lines and use electronic circuitry to select lines. For this mode of operation, if a single picture element in a line changed, the whole line would have to be rewritten. In the strain-biased ferpic, the writing light beam can address and change a single picture element at a time. This is expected to be a particularly important feature when the device is used in some display applications.

5. *Performance Capabilities of Experimental Devices*

a. Sensitivity and Speed. The experimental devices using PVK as the photoconductive film require light fluxes of about 10 mW/cm² (for white

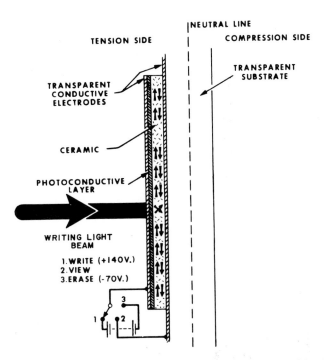

Fig. 16. Selective writing and erasing using a laser beam on strain-biased ferpic.

light from a xenon lamp) for exposure times of ~1 s in order to write an image into the ceramic. This relatively large energy density and slowness of response are attributable solely to the photoconductive characteristics of the PVK. Measurements (Maldonado and Meitzler, 1970) of light-gate structures indicate that domain switching processes can take place in the ceramic in times of the order of microseconds. Thus, the ceramic material itself is capable of operation at speeds several orders of magnitude faster than our existing experimental devices; but questions still remain of whether large enough amounts of polarization can be switched fast enough with reasonable voltages and without causing the sample to explode.

A real-time animated display, laser addressed, operating at TV rates was constructed by Melchior et al. (1970) using a ferpic with CdS photoconductive films. However, the resolution of the device, operated at TV rates, deteriorated considerably from the static value. For TV rate applications the Titus tube described by Marie (1967) is a more attractive device. Experiments have shown that ferpics are suitable for slow scan applications; e.g., as a remote blackboard system in which the moving stylus of existing projection systems is replaced by a laser beam scanning a ferpic. No appreciable change in the resolution capabilities have been observed when writing with a scanned laser beam at 1 ms/cm on ferpics having a CdS photoconductive film.

b. Resolution.   i. Basic resolution capabilities of the ceramic.   The rudimentary ferpic device structure described by Meitzler *et al.* (1970) provides an excellent means of establishing the resolution capabilities of the ceramic. To date, the best resolutions obtained have been slightly in excess of 50 lines/mm in ceramic plates 50-μm thick. At this resolution limit, a complete cycle of light to dark variation takes place in a distance of 20 μm. Since the grain size in this material is 1–2 μm, it is not expected to limit the resolution of any practical device. Instead, fringing fields in the ceramic are considered to impose the basic limitation on resolution: note that with plate thicknesses of 50 μm, the 50 lines/mm already observed corresponds to 2–3 cycles of variation over a distance equal to the plate thickness.

An example of the image storage capabilities of the device is shown in Fig. 17 for a device using a CdS photoconductive film on a PLZT 7/65/35 ceramic plate 75-μm thick. The image of the resolution test chart was

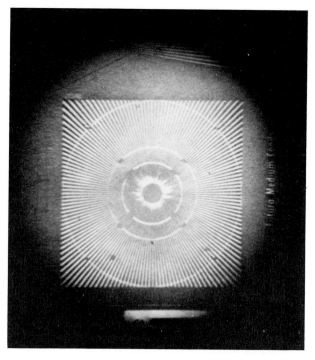

Fig. 17.   Stored resolution test chart in strain-biased ferpic.

stored using a contact printing technique. The writing light was of 514-nm wavelength (filtered from a xenon lamp) and illumination of 10 mW/cm². A 75-V, 16-ms-long pulse was applied to the device during writing. The contrast ratio was about 9 dB and was mainly limited by the depolarization of the light in the ceramic plate (as will be discussed later). The resolution of the device was over 40 lines/mm and was limited by the plate thickness ($\sim$75 μm).

ii. Anisotropic resolution. Strain-biased ferpics exhibit anisotropic resolution (Maldonado and Meitzler, 1971). A series of experimental runs has been made to determine quantitatively how the resolution capability of the strain-biased ferpics depends upon the thickness of the ceramic plate. These tests were made using the PLZT 8/65/35, high-transparency material. Figure 18 shows the data obtained for resolution as a function of plate thickness. The upper curve shows the resolution for test lines perpendicular to the tension axis as the plate thickness increases. The two curves converge as the plate becomes thinner, so that a plate $\sim$50 μm thick has nearly isotropic resolution. (However, even though the measured resolutions in the two directions were nearly equal, we noted in the course of performing these tests that the difference in contrast was still noticeable.)

The data of Fig. 18 apply specifically to the case of devices operated with S_{33}, a tensile strain. As has already been indicated, resolution is influenced not only by the thickness of the plate, but also by the detailed nature of the strain state induced by the applied stress. For example, the device operated with compression along x_3 was more isotropic in resolution than the device operated in tension. The most uniform characteristic has been obtained by clamping the Plexiglas in a small vise and applying a compressive stress along x_2 without flexing the substrate.

A quantitative theory for the observed anisotropy in resolution has not been developed. However, our experimental work strongly suggests that the effect can be minimized by optimizing the combination of sample geometry and strain biasing.

iii. Effects of material inhomogeneities. The resolution in the images shown in Fig. 17 is affected by micrometer sized particulate inhomogeneities and by index of refraction inhomogeneities that are microscopic in extent and occur in the fabrication of the material. With the advent of the new fabrication techniques (Haertling, 1971) for the high lanthanum composition materials, these defects are virtually eliminated.

c. Optical Transmission. The optical transmission characteristics, from 400 nm to 700 nm, of three complete devices (including Plexiglas plate, epoxy-bonding layer, indium tin oxide film, ceramic plate, PVK film

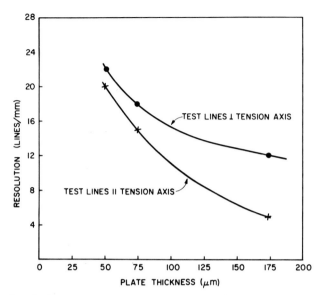

Fig. 18. Resolution versus plate thickness for strain-biased devices using PLZT
8/65/35. All measurements performed with sample under tension ($\sim 3 \times 10^{-3}$ strain).

and semitransparent Au film) are shown in Fig. 19. The optical transmission for a PVK film on Plexiglas is also shown in Fig. 19. The Plexiglas
plate by itself was found to transmit about 93% from 350 to 1000 nm. We
observe from the figure that the wavelength dependence of the transmission curves is mainly determined by the PVK film. We also observe that,
for a given wavelength, the transmission is strongly dependent upon the
lanthanum concentration of the ceramic plate, as well as on the plate
thickness. The device made with a 37-μm-thick plate of 7% La material
has only slightly less transmission at 600 nm than a device made with a
75-μm-thick plate of 8% La material at the same wavelength. The device
with the 75-μm-thick plate of 2% La material has much less transmission
than the other two. The 2% La device has approximately 10% transmission at 600 nm.

The 7% La device has, at 600 nm, approximately 7 dB of optical insertion loss. Of this total, less than 1 dB is lost in the ceramic. The remainder
is divided between the Au film (about 3 dB) and the PVK film (about
3 dB). The use of the Au film in contact with the PVK is necessitated by
the inability of the bonded ceramic plates to withstand the temperatures
involved in sputter depositing an In_2O_3 transparent electrode. The difference between the expansion coefficients of the ceramic and substrate causes

breaking of the ceramic plates. Devices have been made with indium tin oxide films deposited on films of CdS deposited prior to the bending operation.

   d. Brightness and Contrast Ratio of the Projected Image.   The brightness of the projected image, for a given light source, is determined by the optical transmission of the layers making up the complete device (as discussed above) and by the maximum achievable change in retardation. The fact that less than a half-wavelength change in phase retardation is obtained in the present devices results in reduced effective transmission. This results because in order to achieve maximum contrast we cross the polarizer and analyzer and adjust the compensator for best extinction in the dark areas of the picture. With less than $\lambda/2$ retardation, the highlights then correspond to less than the maximum available brightness. This loss could be reduced by allowing the light to pass twice through the device (as discussed below).

   Using white light from a 1000-W xenon lamp with a strain-biased ferpic (made with a 37-μm-thick plate of 7/65/35 La, operated under tension and positioned between crossed polarizers), we measured a highlight illuminance of 35 fc on a screen 5 ft from the analyzer. The projected image covered an area of approximately 1 ft². With a high efficiency screen (e.g., a Kodak Ektalite screen with a gain of 4.5), this illuminance produced a luminance of 157 fL. When a 573-nm monochromatic filter was inserted into the beam, the illuminance was reduced about 10 dB, and the contrast ratio measured for this condition of transmitted light was at least 15 dB.

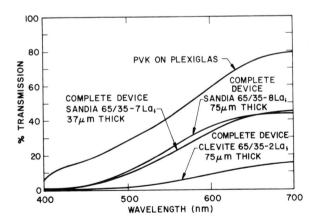

Fig. 19.   Optical transmission characteristics of three complete strain-biased devices. The transmission for a PVK film on Plexiglas is also shown in the figure.

By the use of CdS photoconductive films and indium tin oxide electrodes, the light transmission can be improved (by ∼5 dB).

e. Gray-Scale Capability. Most of the early experimental results were obtained with high-contrast images, but the basic device does have a gray-scale capability. This capability is directly related to the fact that the ceramic can be partially switched to states of polarization intermediate to those of saturated remanence and, hence, intermediate levels of phase retardation can be obtained. The basic mechanism by means of which this takes place during the process of writing in an image is called charge-limited switching. Each element of area in the ceramic is switched by current controlled by the high value of photoconductor resistance effectively in series with the ceramic. The gray-scale capability of the device is shown in Fig. 20.

To obtain the stored picture shown in the photograph, a 35-mm negative was put in contact with the device and was illuminated with white light of about 100 mW/cm² for 2 s.* A PVK photoconductive film was used on a 63-μm-thick 7/65/35 La ceramic plate. The image was stored in the ceramic plate and projected (magnified approximately ten times) on the film

Fig. 20.   Gray-scale capability of the strain-biased devices.

---

* This large light flux was necessary because of the organic photoconductive film (PVK) and was found to depend on the thickness of the film. With CdS photoconductive films, writing times of the order of $\frac{1}{30}$ s were observed.

of a Polaroid camera (with a focal plane shutter and frosted-glass viewer) using monochromatic light of 573-nm wavelength. The grainy appearance of the image is primarily attributable to the surface finish of the ceramic plate.

Gray-scale variation is evident although there is an appreciable compression of the range of variation resulting from the fact that the device tested had less than a half-wavelength range in phase retardation.

B. Reflection Mode Devices

1. *Device Structure*

Strain-biased devices operated in a reflection mode have been constructed (Maldonado and Anderson, 1971). The device structure is shown in Fig. 21. It differs from the structure shown in Fig. 11, by the addition of an array of metallic reflecting dots embedded in an optional opaque film between the ceramic plate and the photoconductive film. The writing and reading operations are shown in the figure.

2. *Device Features*

The reflection mode device has three major advantages relative to the transmission mode devices: (1) better resolution obtained by the use of thinner plates to get a $\lambda/2$ change in retardation because the reading light passes twice through the device; (2) isolation between the reading and writing light provided by the opaque film; and (3) because the projection light does not pass through the conductive film, avoidance of this potential source of light loss. Indeed one has complete freedom of choice of the photoconductor which does not have to be transparent to the projection light at all.

3. *Performance Capabilities*

A picture stored in a reflection mode device is shown in Fig. 22. An array of 250 metal dots per inch was deposited on the ceramic plate. The photoconductive film was PVK and no opaque film was used. The resolution of the device is limited by the array of dots, not by the ceramic plate, as can be seen from Fig. 22.

The change in phase retardation obtained in the device from the ON to the OFF condition was approximately $\lambda/2$ at 576 nm. The low contrast observed in the projected image ($\sim$6 dB) is attributed to the depolarization of the reflected light by scattering in the ceramic plate. This limitation is further discussed below.

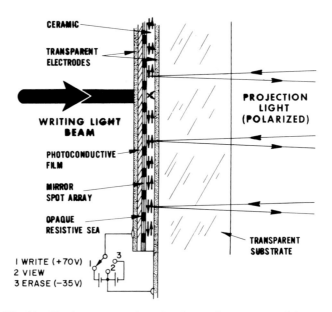

Fig. 21.   Device structure for reflection mode strain-biased device.

The insertion loss of the basic reflection device was measured to be about 4.5 dB, and arises from several sources:

1. The reflection losses for aluminum films are about 1 dB at 576 nm (calculated for normal incidence inside the ceramic plate).

2. The light lost in the gap between reflective dots spacing for the 250 dot per inch pattern are about 2 dB.

3. The absorption losses in the ceramic plate, again at 576 nm, are about 0.01 dB/μm, so that for the 75-μm plate the loss is about 0.75 dB per light pass (1.5 dB total).

In addition, in our experiment, there was a 3-dB loss in the polarizer and a 6-dB loss in the beam splitter used to separate incident and reflected light. The total insertion loss was thus $\sim$13.5 dB. In principle, this could be reduced by 6 dB by the simple expedient of using a polarizing beam splitter (Pritchard, 1969), in which case the total insertion loss would be 7.5 dB.

Although the reflection system avoids photoconductor losses, the total light losses might, in practice, be comparable for both reflection and transmission systems. This is because reflection losses from the metal dots and beam splitting losses can total as much as the loss through present photoconductive films. An equivalent transmission mode system will have about

7.5-dB total insertion loss (assuming 3-dB loss in the photoconductor and transparent electrodes). In the reflection mode case, with careful system design, the beam splitter losses could be reduced to about 4 dB using polarizing prisms (including the 3 dB inherent in the initial polarizing operation). The dots could be made of silver and spaced more closely, reducing the reflection losses to perhaps 1.5 dB maximum. With these assumptions, the total insertion loss of the system would be about 7 dB, which is comparable to that of the transmission system. In practice, devices may have even higher insertion losses, because it is not yet clear whether or not they can be made of plates thick enough to provide a full $\lambda/2$ retardation. This holds equally well for both transmission and reflection mode devices, and is a consequence of our inability, with the residual scattering observed in existing material, to obtain adequate contrast in thick plates. It may thus be necessary to compromise between contrast and insertion loss. Neglecting other sources of loss, the transmission of the ceramic when placed between crossed polarizer and analyzer is given approximately by

$$T = \sin^2\left(\frac{\pi}{2}\frac{\Delta\ell}{\Delta\ell_{1/2}}\right) \tag{2}$$

where $\Delta\ell$ is the actual switched change in the optical path length, and

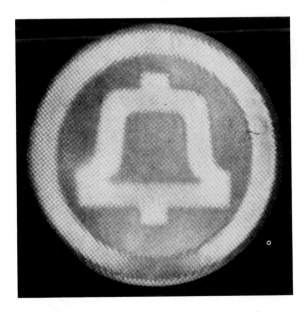

Fig. 22.   Stored image in reflection mode strain-biased device.

$\Delta \ell_{1/2}$ the path length change needed for $\lambda/2$ incremental retardation. Halving the sample thickness would evidently add another 3-dB insertion loss—which might or might not be tolerable. On the other hand, decreasing the thickness by one-third would add less than 1.3 dB and would probably represent a reasonable trade-off, assuming that the light scattering in the ceramic decreases at least linearly with decreasing plate thickness.

C. PERFORMANCE LIMITATIONS IN STRAIN-BIASED DEVICES

Even when the strain-biased devices are illuminated with monochromatic light, it is generally difficult to obtain a large (ON-OFF) contrast ratio (>10 dB) when a large collecting aperture is used. This limitation is due to the depolarization of the reading light in the ceramic plate and is a property of the ceramic material.

To minimize insertion loss, the device designer would like to use a ceramic plate thick enough to produce a $\lambda/2$ change in retardation between the full ON and the OFF states. In general, because of the depolarization of the light in the ceramic plate, some compromise must be made between total insertion loss and ON-OFF ratio (contrast ratio in display devices). However for the 6/65/35 composition, no compromise seems to be necessary. This is shown in Fig. 23 for PLZT 6/65/35 in a polarization switch

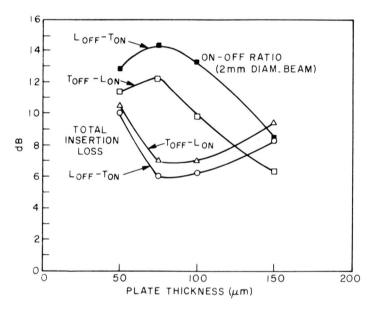

Fig. 23. ON-OFF ratio and insertion loss versus plate thickness for strain-biased devices using PLZT 6/65/35. 5° Aperture (2 mm diam), $S = 2.5 \times 10^{-3}$, $\lambda = 630$ nm.

configuration (Meitzler *et al.* (1971)). We observe from the figure that for this composition at 630 nm a plate thickness of 75 μm corresponds to both the maximum ON-OFF ratio and minimum insertion loss. This fortuitous behavior is due to the combined effect of the increased light depolarization as the ceramic plate thickness is increased, and the increase in the change of retardation with increasing thickness. The above behavior cannot be explained in any simple way by taking account only of the transmission losses in the device and the change in retardation with plate thickness, but is a result of the complex scattering and depolarizing properties of the plates. When the devices are operated with nonmonochromatic light of certain bandwidth, $\Delta\nu$, we expect that the contrast ratio will decrease even further for a given plate thickness. This is due to the fact that in the OFF condition, the retardation is not zero. Therefore, a complete light extinction between crossed polarizers is only possible for a particular wavelength.

Another noteworthy point in Fig. 23 is that it is advantageous to operate the longitudinal mode devices with the maximum birefringence state (L-state) as the OFF condition, and the minimum birefringence state (T-state) as the ON condition. This is because of the higher scattering losses encountered in the L-state.

The contrast limitations of the strain-biased devices can be ameliorated by appropriate choice of the ceramic material and plate thickness, and by appropriate compensation for wide bandwidth operation (for example, by using two identical ceramic plates with equal retardation in the OFF condition and with their optic axes at 90° with respect to each other).

Lifetime studies of PLZT ceramics in the polarization switch structure have shown that the devices can operate for more than 10^8 cycles with no appreciable aging. The lifetime capabilities of the ceramic material have been shown (Fraser and Maldonado, 1970) to be over 10^9 cycles.

We know as a result of attempts to realize a real-time, animated display that switching lifetime seriously degrades as switching speed increases. Laboratory devices operated at TV rates have lasted for times less than one hour. On the other hand, devices switched at rates of 1000 Hz have shown the ability to survive 10^9 cycles.

D. APPLICATION TO PAGE COMPOSERS FOR OPTICAL MEMORIES

C. E. Land of the Sandia Corporation proposed several years ago (Land, 1970) a device consisting of an array of light gates suitable for use as a page composer for an optical memory. The light gates in this device utilize the transverse electrooptic effect. Due in large part to the fact that the electrode configuration was very complicated, and that very high voltages were needed for relatively large apertures, this device has not been developed for use in optical page composers having large numbers of elements. However, some work is presently being conducted in other laboratories to utilize this device with the nonmemory PLZT 9/65/35 materials in page composer applications (Roberts, 1972).

Strain-biased devices offer certain advantages relative to the transverse mode array of light gates in page composer applications. The most important is that large apertures can be obtained with relatively low voltages. Also because of the possibility of X addressing the device from one side of the plate while Y is addressing the opposite side, a relatively simple electrode configuration can be obtained.

A 900 element X-Y electrically addressed array consisting of 30 transparent electrode rows 250-μm wide and 75-μm apart deposited on the bottom of the ceramic plate, and 30 identical transparent electrode columns on top of the ceramic plate has been constructed (Maldonado (1971)). A photograph of the array is shown in Fig. 24. By taking advantage of the voltage controlled switching mode, it is possible to reduce the effect of disturb pulses. This can be done by a method suggested by Coquin which takes advantage of the hysteresis characteristics of the high lanthanum ceramics. In materials with more than 6% lanthanum, the birefringence versus polarization curve is shown schematically in Fig. 25. Some hysteresis is observed by switching the material between $+\langle P_r \rangle$ and $-\langle P_r \rangle$ and reversing the procedure as shown in the figure. The voltages necessary to switch the ceramic plate in a voltage controlled mode are shown in Fig. 25 below the polarization values. The procedure to avoid the half select problem is as follows. The device is first set at point C by applying voltages of $+60$ V and -60 V to all the rows and columns, respectively. Then the device is set at point α by applying and then removing voltages of $+30$ V and -30 V to all rows and columns, respectively. This last step is equivalent to subjecting all the rows in the array to a voltage of $+60$ V and grounding the columns. To address a particular element on the array, a voltage of $+60$ V is applied to the corresponding row and a voltage of -60 V is applied to the corresponding column. This procedure will set the intersecting element at point C following the curve αBC. The remaining elements along the addressed row and column (with all the rest of the rows and columns grounded) will receive $+30$ V on the row and -30 V on the column. This is equivalent to applying $+30$ V on the row with the intersecting columns grounded. Therefore, since all the elements were previously subject to $+60$ V on the row, no further switching will occur with $+30$ V. In this fashion, the information could be written in all the elements. It is not possible, using this technique, to erase a particular

Fig. 24.   Photograph of 30 × 30 array using a strain-biased device.

element in the array (once it is written) without affecting the other elements in the array. Total erasure is necessary after one page is composed. In order for this technique to be effective, the birefringence change between α and α' must be small relative to the total birefringence change, otherwise a large insertion loss will occur. The shape of the hysteresis curve depends on the lanthanum composition of the material and more work is necessary to select the material that maximizes the device performance.

Roberts (1972) has constructed a 1 × 5 page composer prototype using the strain-biasing technique with satisfactory results. (For device performance results, the reader is referred to his paper.)

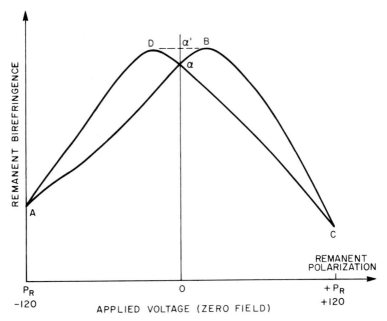

Fig. 25.   Schematic representation of remanent birefringence versus average remanent polarization $\langle P_r \rangle$ for materials with more than 6% lanthanum.

III. Scattering-Mode Image Storage and Display Devices— Cerampics

As mentioned in Section II, C, light traversing a strain-biased plate in the L-state ($\langle P_r \rangle = 0$) is scattered more than when the plate is in the T-state ($\langle P_r \rangle_{max}$). This fact was extended by Smith and Land (1972a,b) to the case of a free ceramic plate (i.e., not bonded). They have shown that when thick plates ($> 125\ \mu m$) of high La composition are used, large changes in the scattering properties of the material are observed between the $\langle P_r \rangle = 0$ and $\langle P_r \rangle_{max}$ states. The original device structure proposed by Smith and Land for the ceramic device is shown in Fig. 26. The device consists of a ferroelectric-photoconductor sandwich structure using one photoconductive film and two transparent electrodes.

Three kinds of devices have been constructed that operate in the scattering mode: (1) the thermally erased device using high lanthanum materials, (2) the thermally erased device using low lanthanum materials, and (3) the electrically erased device. The operation of the thermally depoled device will be described below. The performance of the electrically erased device will be presented in conjunction with the slow scan graphic system.

A. THERMALLY ERASED DEVICES

1. Device Structure

The device structure is shown in Fig. 27 and differs from the basic ceramic structure in that special connections are made to the indium tin oxide transparent electrodes to allow them to be used as electrical heaters for thermally depoling the whole ceramic plate.

2. Thermally Erased Devices Using High Lanthanum Materials

a. *Mode of Operation.* The use of thermal depoling in the operation of the device is based on the observation that for small collection apertures in high lanthanum concentration materials, the total intensity of light scattered by a ceramic plate in the thermally depoled condition is considerably less than the total intensity of light scattered in the electrically depoled condition (zero average remanent polarization obtained by application of electric fields to the ceramic plate).

Storing an image in the device is done after the device is thermally depoled. The writing operation consists of supplying two voltage pulses of opposite polarity to the transparent electrodes. Switching fields are developed in the ferroelectric ceramic only in the illuminated regions. The first pulse is used to switch the average polarization of the region to maximum remanence $\langle P_r \rangle_{\max}$. (The field necessary to pole a region to $\langle P_r \rangle_{\max}$ is about 10 kV/cm.) The second pulse of opposite polarity and smaller amplitude, applied immediately following the first, is used to switch the average polarization to zero ($\langle P_r \rangle = 0$). In this condition, the total intensity of scattered light is a maximum. Therefore, a large contrast ratio from the thermally depoled state can be obtained by placing an aperture

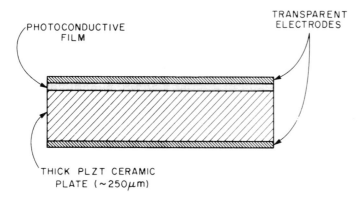

Fig. 26.   Device structure of cerampic device.

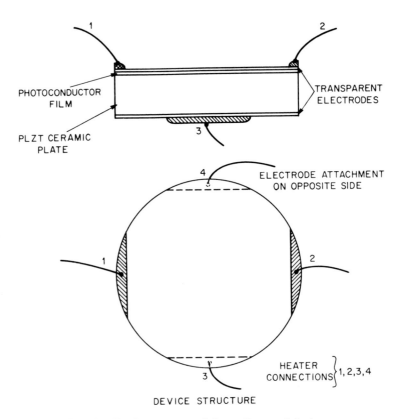

Fig. 27.   Device structure of thermally erased device.

stop in front of the device to form a schlieren projection system and inter-
cept the scattered light.

Erasure is accomplished by means of the two transparent indium tin
oxide electrodes, used as heaters, for which they are connected in parallel
to a current source. As the temperature of the ceramic plate increases,
domain reorientation occurs in a random fashion, the plate becomes micro-
scopically more homogeneous and the total intensity of the scattered light
decreases. This phenomenon continues until a transition point (115°C for
the 7/65/35 material) that is about 30°C below the Curie point is reached.
Above the transition point, the material average polarization is close to
zero. As the plate cools to room temperature, the average polarization
remains zero.

   b. Performance Capabilities.   If a small aperture (larger or equal to the
beam diameter) is placed in front of a thermally depoled device, the total

intensity of the light transmitted through the aperture does not differ very much from the case of a large aperture. This behavior can be seen in Fig. 28 (thermally depoled curve) for the case of a 250-μm plate of PLZT 8/65/35. For a poled sample ($\langle P_r \rangle_{max}$ curve), the same figure shows that the total intensity of the light transmitted through the aperture decreases appreciably as the half-angle aperture is decreased. In a schlieren projection system with a half-angle aperture of 3° ($f/9$) the difference between the thermally depoled state and the $\langle P_r \rangle_{max}$ state is about 3 dB. This means a decrease in insertion loss of 3 dB relative to the electrically erased device. The above also implies that if a portion of the plate is switched to $\langle P_r \rangle_{max}$ from the $\langle P_r \rangle = 0$ state, the contrast of that portion is reduced to only 3 dB (relative to the thermally depoled state) and, therefore, the particular portion has been partially erased. This could be made the basis of a selective erase capability which, while not providing perfect erasure, would nonetheless provide enough contrast with unerased material to be useful.

As described above, the device has two states of light transmission: ON (thermally depoled) or OFF (large angle scattering). The device also has gray-scale capabilities because of the partial switching provided by the illuminated photoconductive film.

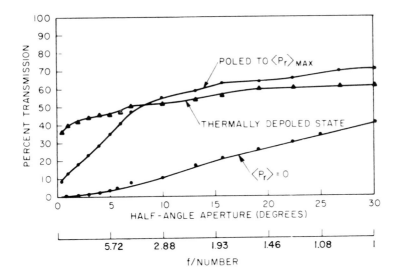

Fig. 28. Light transmission versus detector half-angle aperture (system f number) for a PLZT 8/65/35 device for three different polarization conditions. 250-μm-thick sample, $\lambda = 5682$ Å.

c. *Contrast Ratio and Insertion Loss.*   In order to study the fundamental contrast ratio of the device without the added complexity of having to expose the photoconductor, a light-gate device with no photoconductor was constructed. The device was made with a 250-μm plate of PLZT 8/65/35 with In_2O_3 electrodes on both sides. Figure 29 shows the contrast ratio obtained with this device as a function of half-angle aperture for three different wavelengths.

The figure shows that, for a collection half-angle of 3°, the observed contrast ratio using collimated light is more than 15:1 for the least favorable wavelength (632.8 nm). The transmission for the same device is shown in Fig. 30 for the corresponding wavelengths. We note from the figure that for the 568.2-nm (yellow) line the transmission is about 45% at 3° half-angle aperture. This value of transmission includes the reflection losses ($\sim$1.5 dB) on both plate surfaces, probably partially compensated by the indium tin oxide films. In an actual device, the photoconductor would add about 2 dB to the total insertion loss; therefore, the total insertion loss in an antireflection-coated device would be about 3.5 dB for a 250-μm device with a 10:1 contrast. The figure also shows that the trans-

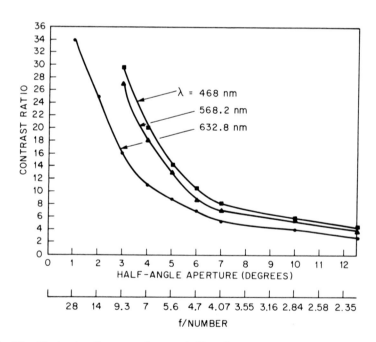

Fig. 29.   Contrast ratio versus detector half-angle aperture for three different wavelengths for a thermally erased device PLZT 8/65/35, 250-μm-thick.

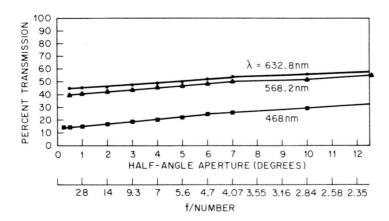

Fig. 30. Light transmission versus detector half-angle for three different wavelengths for the device of Fig. 29.

mission for blue light is about 20% for the 3° aperture. This fact and the transmission characteristics of the photoconductive film (which also transmits less blue) imparts a yellow color to the displayed image. The transmission of the device can be enhanced somewhat by reducing the plate thickness; this in turn would produce a decrease in contrast for a given aperture.

The percent transmission versus wavelength for an actual device (125-μm-thick ceramic plate) with a 4-μm layer of $Cd_{1-x}Zn_xS$ is shown in Fig. 31. The figure shows the transmission for two photoconductors, one with maximum sensitivity at 441.6 nm and the other peaking at 488 nm. The data were taken with the ceramic in both cases in the thermally depoled state using a system with an $f/22$ collecting lens. For lower f numbers, the measured transmission is slightly better. No antireflection coatings are used in our devices; however, some reduction in reflection loss is provided by the index matching properties of the transparent electrodes ($\sim$200-nm thick).

d. Resolution Capabilities. The observed resolution in devices made using PVK and CdS photoconductive films is surprisingly high (about 32 line pairs/mm for the 250-μm plate) corresponding to 8 line pairs per plate thickness. This fact is not well understood, but it has also been observed by Smith and Land in their device. They have attributed this effect to surface deformations that occur on the ceramic plate, due to strain relieving (Land and Smith, 1973).

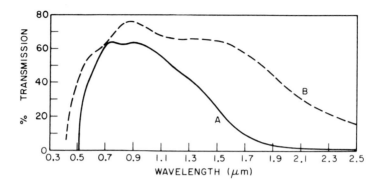

Fig. 31. Light transmission versus wavelength for 125-μm-thick ceramic plate with a 4-μm layer of $Cd_{1-x}Zn_xS$. The cases $x = 0$ (A) and $x = 0.5$ (B) are shown.

e. Summary. The main advantage of the thermally erased device described in this section relative to the electrically erased device lies in the relatively low insertion loss obtained for small apertures. This is very important if the device must compete with equivalent liquid crystal systems. The accompanying disadvantages are that the erasure time is relatively long (seconds) and that the selective erase capability is not complete.

3. *Thermally Erased Devices Using the Low Lanthanum Materials*

The ceramic materials with less than 6% lanthanum have different scattering characteristics than the ones with more than 6% lanthanum (Maldonado and O'Bryan, 1973). This can be seen in Fig. 32 for the 1/65/35 composition.

This figure shows that the material in the thermally depoled state transmits less light than the poled material for any detector half-angle, θ. This behavior is expected from the differences in domain distribution between the thermally depoled and poled states. In the thermally depoled state the domains are randomly oriented, and therefore we expect more light scattering than in the poled state. In this latter state, the domains are predominantly aligned along the poling direction and smaller index of refraction variations are encountered by the transmitted light.

The light-scattering behavior of the materials with the low lanthanum composition shown in Fig. 32 suggest the possible operation of ferroelectric-photoconductor display devices having the scattering condition as the thermally depoled state and the transmitting condition as the electrically poled state. In a conventional schlieren system using an aperture, this mode of operation would provide a displayed image with a dark background and bright characters. We have constructed devices using this mode of operation with satisfactory results in a slow scan graphic display system similar to the one described in Section III. B. The device structure is the same as the thermally depoled device shown in Fig. 27. The indium tin oxide transparent electrodes are used as heaters to bring the material above the transition temperature. (Erasure takes a few seconds when performed this way.) This kind of device has some advantages relative to the thermally depoled devices with high lanthanum ceramics using the reverse mode of operation (in which the thermally depoled state is the transmitting condition) described in Section III, A. One advantage lies in the fact that low f number (and hence bright) systems could be used with relatively high contrast and low insertion loss. For example, Fig. 32 shows that for a half-angle aperture of 20° ($\sim f/1.4$) the contrast ratio between the thermally depoled state and the electrically poled state is about 6:1. The figure also shows that the light transmission of the device $\lambda = 568.2$ nm for an $f/1.4$ ($\theta/2 = 20°$) system is about 58%. This kind of device also offers the possibility of obtaining higher resolution than the scattering devices described in the past. This is due to the fact that the large angular distribution of scattered light obtained in the thermally depoled state allows for the use of thinner ceramic plates to obtain reasonable contrast ratios with relatively low f number systems. These devices may prove suitable in some relatively high resolution (1000 lpi) slow scan applications in which a selective erase feature is not needed.

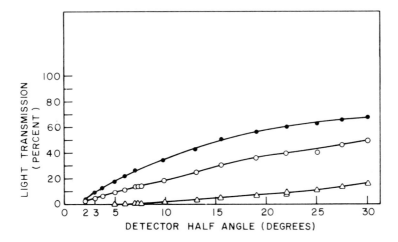

Fig. 32.   Light transmission versus detector half-angle for the PLZT 1/65/35 composition. ●, Poled $\langle P_r \rangle_{max}$, $\lambda = 568.2$ nm; ○, poled $\langle P_r \rangle_{max}$, $\lambda = 514.5$ nm; △, thermally depoled above Curie point, $\lambda = 568.2$ nm.

B. Electrically Erased Device

As mentioned before, because of the different properties of the thermally depoled and electrically poled states, it is possible to operate the device in two different modes: one in which the initial high transmission state is achieved by thermally depoling, the other in which it is achieved by electrically switching the average remanent polarization to its maximum value $\langle P_r \rangle_{max}$. In either mode, the maximum scattering state is arrived by electrically switching the plate to zero average polarization ($\langle P_r \rangle = 0$). As can be seen from Fig. 28, the electrically erased device can operate with low f number systems (f number <4) with reasonable contrast and low insertion loss. Therefore, if contrast ratios larger than about 7:1 are not needed, the electrically erased device is preferred over the thermally erased device for its simplicity of structure (exploded view shown in Fig. 33) and operation.

1. *Performance Capabilities in a Slow Scan Graphics System*

A slow scan system described by Maldonado and Fraser (1973) is shown in Fig. 34. The device is optically addressed with an XY deflected beam from a He–Cd laser. The laser power required at the device can be as low as 250 μW for devices with only one photoconductor. For symmetric devices, the required power is close to 1 mW (some fraction of the light is absorbed in the first addressed photoconductor). The device is first prepoled to maximum remanent polarization by applying a field of about 10 kV/cm to the transparent electrodes while the photoconductor is illuminated with white light from the projection lamp. The writing operation is done by applying pulses of opposite polarity to the one used for prepoling and simultaneously addressing the device with the focused laser beam. The applied pulse amplitude and duration are set to switch the remanent polarization to zero in the addressed regions. The switched regions look black on the screen because they scatter light out of the aperture of the projection system. By proper combination of the width and amplitude of the applied pulses, projection of the image can take place simultaneously with write-in, without any appreciable fogging of the storage medium by the viewing light. This point will be discussed in more detail later. The repetition rate of the applied pulses is about 500 Hz. This repetition rate is fast enough to reproduce analog information, such as ordinary handwriting, on a conventional blackboard (McDowell and O'Boyle, 1971).

The contrast of lines written with the laser beam decreases as the linewidth is made narrower than the plate thickness. However, even for a single photoconductive film structure, reasonable contrast is obtained with

EXPLODED VIEW

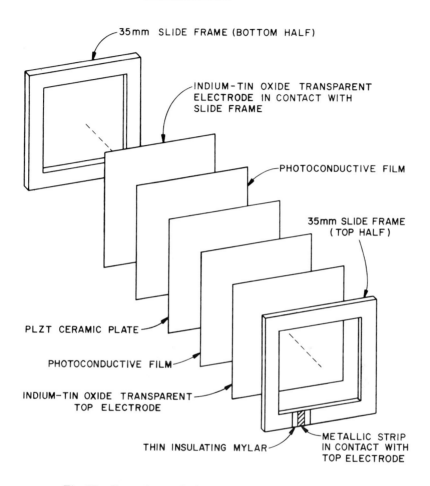

Fig. 33.   Scattering mode devices using a symmetric structure.

linewidths about one-half to one-third the plate thickness (which is typically $\sim$250 μm). For symmetrical devices, the resolution is about a factor
of two better. After the information is written and stored in the ceramic
plate, it can be either totally or selectively erased. The former is accomplished by illuminating the whole plate with white light while applying the
poling voltage for a short time (typically a few milliseconds for conventional xenon arc lamps of 150 W). Selective erasure is achieved by illumi-

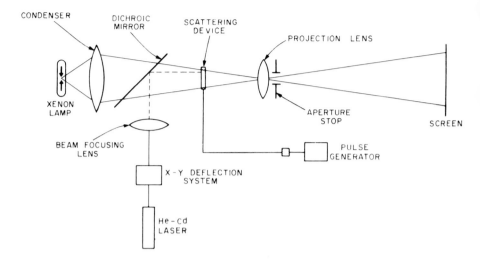

Fig. 34.   Slow scan graphics system.

nating the desired region with the laser beam and simultaneously applying pulses of the same duration as the ones used for writing ($\sim$5 kV/cm), but with opposite polarity and larger amplitude (10 kV/cm). An example of selective erasure is shown in Fig. 35. After a portion of the stored image is selectively erased, new information can be written in the previously erased region.

Because of the large light scattering angle obtained with thick ($>$125 μm) plates of PLZT ceramic, it is possible to obtain reasonable contrasts (5:1) even when the device is used in a large aperture projection system (typically about $f/3.5$ for a 35-mm slide projector). This fact is very important to the system designer because a bright, small aperture (f number $>f/5$) projection system is expensive due to the need for high-power arc lamps with small arc gaps.

The color of the projected image is yellow in a high f number system due to the transmission characteristics of the photoconductive films and the scattering properties of the ceramic material. (Scattering increases with shorter λ so the blue light is scattered out of the system aperture). However, in a low f number system, the main contribution to the color comes from the photoconductor, and the projected image has only a slight yellow tint that will not be objectionable in most applications. There may also be some coloration due to the dichroic mirror. This can be minimized by proper mirror design.

2. Summary

The main features of the display device described here are as follows: (1) It is an all solid-state display device. (2) The storage medium is thin and self-supporting, and can be prepared so that previously written slides can be viewed in any conventional 35-mm projector. (3) The photoconductive film used in the device permits reading and writing to be done simultaneously. (4) The photoconductor sensitivity can be tailored for the He–Cd laser wavelength (441.6 nm) which is potentially an inexpensive source of writing light. (5) The device is capable of operation with very low laser power (<1 mW). (6) The device has selective erase capabilities.

Abbreviated statements of these six advantages were written into a ceramic display device and projected onto a photographic film. Figure 36 is a reproduction of the photograph obtained in this manner. (The writing on the original device occupied an area ∼1.5 cm².)

C. INFLUENCE OF MATERIAL PREPARATION ON DEVICE PERFORMANCE

1. The Photoconductive Films

Initially, devices fabricated with a photoconductive film sensitive to the 441.6-nm line of the He–Cd laser were characteristically high in insertion

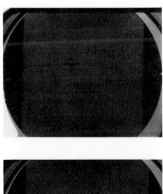

Fig. 35. Selective erase capabilities of scattering-mode device.

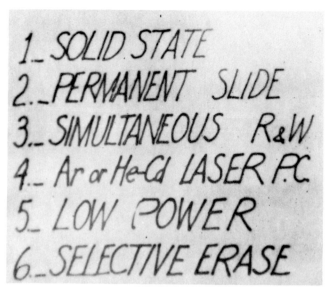

Fig. 36.   Features of the scattering-mode ferroelectric display device written by a laser beam on a 250-μm-thick device.

loss and low in photosensitivity compared to devices fabricated to be used with the 488.0-nm line of the argon laser. The differences in performance are shown in Fig. 37 and Fig. 38 for two devices (using the symmetric structure) made from the same ceramic boule. One device has the 441.6-nm photoconductor and the other device has the 488-nm photoconductor. Figure 37 shows the device's contrast ratio versus detector half-angle (system f number). The data were taken with the 632.8-nm line of a He–Ne laser. Figure 38 shows the insertion loss versus detector half-angle aperture (system f number) taken at 632.8 nm. We observe from the figure, that the device made with the 488-nm photoconductor has higher contrast ratio and lower insertion loss than the 441.6-nm device. The behavior of the insertion loss is anomalous because the transmission of the 488-nm films is lower than the 441.6-nm films at 632.8 nm, as was shown in Fig. 31. The reason for the observed difference in total device performance seems to be the presence of different amounts of strain in the ceramic plates due to the different mechanical behavior of the two photoconductive films (488 nm and 441.6 nm) on the ceramic plate. A permanent strain in the material causes some of the domains to remain parallel to the plate surface even when the device is switched to the ON condition. This increases the insertion loss and lowers the contrast for a given detector angle. The following experimental results are shown to confirm this dependence of device performance on different surface treatments. Figure 39 shows the difference in contrast ratio versus detector half-angle (system f number) for two different thin films deposited on the ceramic plate. One film is evaporated Cr–Au electrodes (<100 Å thick); the other film is conventional indium tin oxide sputtered film ($\sim$200-nm thick). The indium oxide film was deposited on the same ceramic plate after the Cr–Au film was removed by

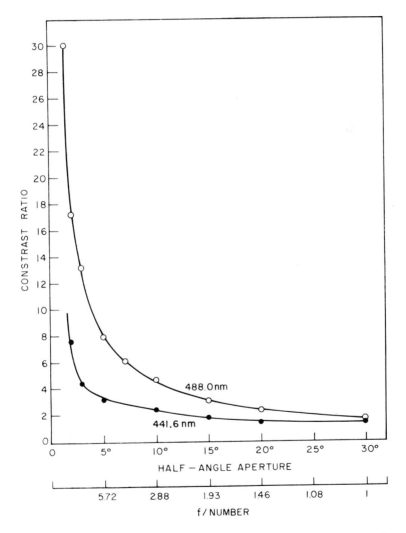

Fig. 37. Contrast ratio versus detector half-angle (system f number) for two devices with 441.6 nm and 488 nm peak response. The data were taken at 632.8 nm.

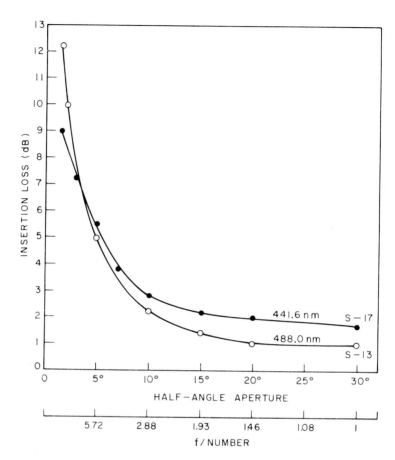

Fig. 38.   Insertion loss versus detector half-angle (system f number) taken at 632.8 nm.

gently polishing, and after the ceramic plate was thermally annealed. We observe from the figure that the device with indium tin oxide electrodes has better contrast ratio than the device with Cr–Au electrodes. The surface polishing after the removal of the Cr–Au film (done with 1 μm diamond powder) was not as good as the initial surface polishing before the Cr–Au electrodes were deposited (done with $\frac{1}{4}$ μm diamond). After considerable development of the photoconductor deposition techniques used for the 441.6-nm sensitive layers a lower strain in the ceramic plate were achieved and a level of device performance comparable to that of the 488.0 nm devices resulted.

2. *The Ceramic Materials*

The light scattering in PLZT ceramics results from the index of refraction discontinuities at domain boundaries (Coquin, 1972). A light beam traversing a 250-μm-thick ceramic plate undergoes a number of scattering events since the average scattering center appears to be 2–5 μm in diameter. Scattering at domain boundaries results in angular dispersion of a collimated incident beam and in depolarization of a polarized incident beam. For device applications, it is important to have (1) a broad angular distribution of the output light from any region in the ceramic plate in the electrically depoled state, and (2) a narrow angular distribution of the output light from regions in the poled condition. Those two conditions are not easily achieved in practice for all ceramic compositions and grain sizes. The two conditions mentioned above depend on the relative domain configuration between the poled and the electrically depoled states. Therefore, they are expected to depend, for a given composition, on the material grain size which determines the size and number of domains in the two states.

Experimental results presented in this section show the dependence of the contrast ratio on grain size and wavelength. The contrast ratio measured in these experiments is the ratio of total light entering the detector with the device in the electrically poled and depoled conditions, respectively. The data were obtained using a PLZT 8/65/35 composition and

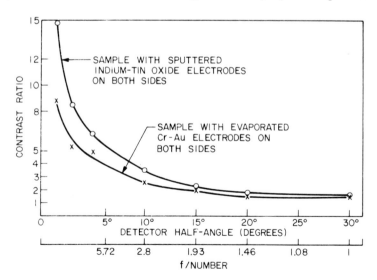

Fig. 39.   Contrast ratio versus detector half-angle (system *f* number) for two different thin films deposited on the ceramic plate.

probably are valid only for that particular composition. The materials were prepared by O'Bryan and Thomson of Bell Laboratories using the same initial proportion of ingredients for all samples (only the hot-pressing temperature was varied to change the grain size from 2.8 to 6.8 μm). Therefore, we expect that the final compositions were the same, and the experimental results were dependent upon grain size variations only. The uncertainty regarding the device surface roughness for scattering measurements was resolved by polishing all the samples of different grain size at the same time. The final thickness was 250 μm. Indium tin oxide transparent electrodes were deposited on both sides of the plates and therefore the complications associated with the presence of photoconductive films were avoided.

Figure 40 shows the contrast ratio versus grain size at 632.8 nm using a detector half-angle of 2°. (Essentially the same results are observed for larger detector apertures.) A maximum in contrast ratio is observed between 4 and 6 μm. This result has been confirmed by many different devices; our best devices have been made with materials of grain size of 4–5 μm. The wavelength dependence of the contrast ratio for several grain sizes is shown in Fig. 41 for a detector half-angle of 2°. We observe from the figure that a well-defined maximum in contrast ratio is observed for all grain sizes at some wavelength. In particular, for the 2.8 μm sample, a sharp maximum is observed. When the detector half-angle is increased to 10°, this sharp maximum broadens and the maxima for other grain sizes disappear entirely (as can be observed in Fig. 42). We also observe that the contrast ratio decreases as the wavelength increases; in particular, for the 4.7-μm sample, the contrast ratio decreases by about a factor of 2, from 440 to 630 nm. The sharp maximum in contrast ratio versus λ (shown in Fig. 11) for the 2.8-μm sample cannot be accounted for completely by using either the light-scattering theory developed by Coquin or the analysis of Delissa and Seymour (1973).

3. Lifetime of Scattering-Mode Devices

As mentioned in the section on strain-biased devices, the lifetime of the ceramic materials at low switching rates can be larger than 10^8 cycles. The structure used to determine the lifetime is very similar to the cerampic structure except that no photoconductive film was employed. However, the main limitation in the lifetime of the devices is not fatigue in the switchable polarization of the ceramic material. The main causes of failure are cracks that develop in the plates when the whole active area of the plate is switched. This failure may be caused by microcracks introduced during material processing, e.g., during the cutting and polishing opera-

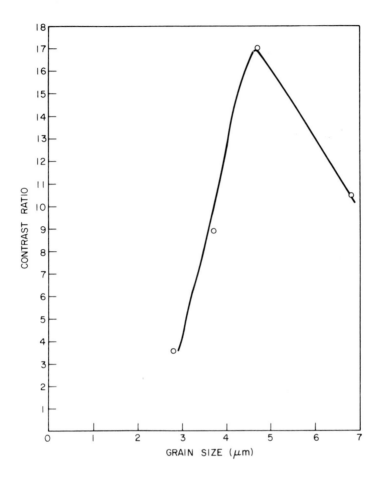

Fig. 40.   Contrast ratio versus grain size at 632.8 using a detector half-angle of 2°.

tions. Also, defects in the ceramic materials (voids, etc.) may cause weak regions where cracks can develop under the strains produced during switching. Cracks have also been observed developing in regions where the ceramic plate was pressed nonuniformly in the slide holder. To minimize the probability of cracking, we operate the devices in a manner that avoids switching the whole plate through the electrically depoled state. The polarization is switched between $\langle P_r \rangle_{\max}$ and $\langle P_r \rangle = 0$ using the same set of polarity voltages every time. A whole plate is never switched between $+\langle P_r \rangle_{\max}$ and $-\langle P_r \rangle_{\max}$ except a few times (very slowly) when the plate is operated for the first time. (In other words, the polarization is only fully saturated in one direction.) This initial cycling was found to minimize any tendency for the device to retain an image after being erased. Another precaution used in the operation of the devices is never to apply an electric field to plates made with 7/65/35 ceramic larger than 10 kV/cm. With improved fabrication techniques, we presently have devices that have operated for a long time. Failure (cracking) normally occurs in the first few hours of operation, so that if a device survives that initial period, the lifetime will be determined by the fatigue in the material ($> 10^8$ cycles).

D. HIGH-SPEED OPERATION

Although we have reported only slow scan applications of the scattering-mode devices, these devices are capable of high-speed operation. Work recently done by DiSabato and Fraser (1973) has shown operation with pulses less than 10 μs with laser power of $\sim$20 mW. The resolution obtained with short pulses was the same as the resolution obtained under slow scan conditions; however, the contrast for an $f/3.5$ system was about 3:1. Because of cracking of the plates, it is expected that only writing of small areas could be done at less than 10 μs. Erasure of the whole plate will probably have to be done at slower rates to avoid cracking.

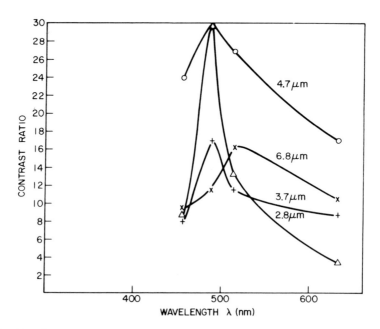

Fig. 41. Contrast ratio versus wavelength for several grain sizes for a detector half-angle of 2°.

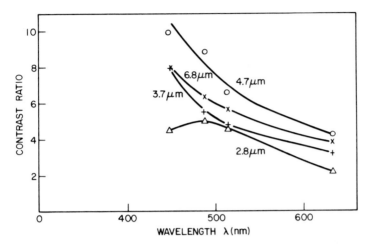

Fig. 42. Same as Fig. 41 for a detector half-angle of 10°.

IV. Summary of the Performance of Display Devices

Table I summarizes the performance of the different types of ferroelec-tric-photoconductor display devices. The performance was normalized to a 10:1 contrast ratio for all the devices.

V. Alphanumeric Display Devices

A. Longitudinal Mode, Scattering Devices

These devices make use of the field induced phase transition occurring in the PLZT 9/65/35 materials. These materials are penferroelectric at room temperature. Therefore, they are very transparent with very low light scattering above 15°C. When an electric field is applied through the thickness in relatively thick ceramic plates, the material becomes ferro-electric and light traversing the plate is depolarized. When the field is removed, the plate returns to the nonscattering state. This effect has been used by Haertling and McCambell (1972) and Smith and Land (1972b) in alphanumeric devices using a ceramic plate, in which a seven bar elec-trode pattern was deposited on one side and a transparent electrode on the opposite side of the plate. The plate was located between crossed polarizers and a mirror plate was set below the lower polarizer to make an ambient light reflection structure. With no applied field the light reflected is deter-mined by the extinction of the crossed polarizers. When the electric field is

TABLE I: Performance for a 10:1 Contrast Ratio

	Approximate maximum working voltage	System insertion loss (dB)	Resolution (lines/mm)	Plate thickness (μm)	Selective erase capability	Features
Strain-biased transmission mode	75	8.5	40	75	yes	Rigid construction, multicolor operation, relatively low voltages
Strain-biased reflection mode	37	8.5	80	37	yes	Same as above plus isolation between reading and writing, higher resolution (thinner plates), allows use of nontransparent photoconductors, color of projected image does not depend on transmission characteristics of photoconductor
Transmission scattering mode (could also be operated in a reflection mode)	250	7.3	30	250	yes	Simpler construction, thicker plates
Transmission thermal erased scattering mode (could also be operated in a reflection mode)	175	<3	40	~175	yes (but not complete; ~3 dB contrast ratio on erased image)	Same as above, low insertion loss

applied to the bars across the ceramic plate, the light is depolarized and the addressed bars transmit light.

1. Device Performance

The device performance is limited by the inherent insertion loss of this mode of operation. For example, an ideal device (properly antireflection coated) in which an ideal polarizer, an ideal analyzer, and an ideal mirror are used will have an insertion loss for ambient light reflection of 9 dB (6 dB for transmission). In practice, the insertion loss will be higher due to the losses in the polarizer sheets, mirror, antireflection coatings, etc. Practical devices have insertion losses of the order of 14 dB in reflection and 9 dB in transmission. For operation in transmission, this insertion loss might be acceptable if enough power is available for the light source. For operation in ambient light reflection, the insertion loss is too high for most applications (except perhaps where high levels of ambient illumination are always available). Voltages of the order of 200 V are needed for the operation of this device.

The above drawbacks have to be weighed against the relatively low power consumption of the device and its simplicity of construction.

Another longitudinal mode device considered at Bell Laboratories was of the birefringence type and used the strain-biasing technique to take advantage of the longitudinal changes in birefringence of the bonded ceramic plates. It was found that the contrast ratio of this device was limited by the residual light scattering of the ceramic materials under strain, and work along those lines was discontinued.

B. TRANSVERSE MODE, INTERDIGITAL-ARRAY, BIREFRINGENCE DEVICES

The results given in this section are very preliminary and will be presented for the sake of completeness. In order to reduce the operating voltages and improve the performance, the devices described in this section operate in a transverse electrooptic mode and utilize an interdigital array of electrodes for each of the bars of the seven bar digit on the surface of a low-scattering 9/65/35 ceramic plate. The combination of interdigital electrode arrays and the electrooptic properties at 9/65/35 ceramic has been used before in goggles to prevent flash blindness (Harris and Cutchen, 1972) and in an electronically variable, neutral density filter. Further refinements of the electrode configurations and materials lend promise for devices truly useful as numeric displays.

1. Device Structures and Performance

All the devices described herein operate with ambient light in reflection (operation in transmission is also possible). For operation in the reflection mode, the ceramic plate is located between one polarizer and a diffusing mirror. The mirror can be deposited on one side of the ceramic plate (as mentioned below). In this fashion, black numbers will be observed in a white background with a $\lambda/4$ retardation in the ceramic plate. If white numbers are required on a black background, a $\lambda/4$ plastic sheet could be located between the polarizer and the ceramic plate. However, the viewing angle of the device will be reduced somewhat due to the relatively low index of refraction of the $\lambda/4$ plastic sheet.

a. Devices with Interdigital Arrays in Registration. This device is shown in Fig. 43. It consists of a ceramic plate in which overlapping interdigital array electrodes in registration are deposited on both sides of the plate* forming the bars of a seven bar digit as shown in Fig. 44. After this operation, a TiO_2 layer is deposited on one side of the plate, to provide electrical isolation and index matching, followed by a diffusing metallic mirror (or reflecting thin film layers). The other side of the plate is antireflection-coated with an aluminum oxide film.

Figure 45 shows the retardation Γ versus applied electric field E in a 100-μm-thick plate of PLZT 9/65/35 for various gap (d) to thickness (t) ratios. The $\lambda/4$ voltage versus plate thickness for a 50-μm gap is shown in Fig. 46. We observe that a 75-μm plate needs about 32 V for 140 nm of retardation. We also observe from Fig. 45 that due to the nonlinear behavior of Γ versus E, the device might be suitable for matrix addressed

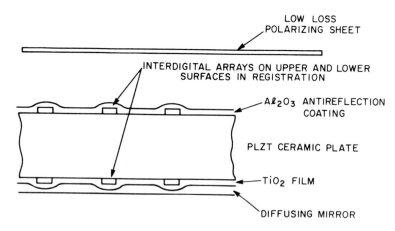

LOW LOSS
POLARIZING SHEET

INTERDIGITAL ARRAYS ON UPPER AND LOWER
SURFACES IN REGISTRATION

Aℓ_2O$_3$ ANTIREFLECTION
COATING

PLZT CERAMIC PLATE

TiO$_2$ FILM

DIFFUSING MIRROR

Fig. 43.   Device structure for devices with interdigital arrays in registration.

* If voltages of the order of 60 V are available, the array may be deposited on only one side of the plate.

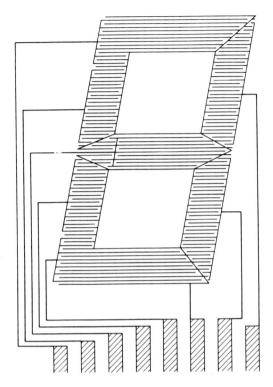

Fig. 44.   Seven bar digit using interdigital arrays.

displays, although the electrode configuration could be very complicated. Recently, a new material has been developed by H. M. O'Bryan, Jr. (Private communication) in response to the needs of this application that has a quadratic electrooptic effect about two times larger than the material used for Fig. 46. With this new material, operation with about 20 V in a 50-μm gap is possible if wide electrodes are used. (This fact will be discussed in Section V, 4.) It is expected that materials with even larger quadratic electrooptic effects can be made using the same techniques, therefore reducing the operating voltage even further. It seems possible, with the material used in the data of Fig. 47, to reduce the operating voltage to about 20 V by reducing the electrode gap to about 25 μm for a 50-μm-thick plate. The dashed curve of Fig. 47 was obtained by taking the curve for $d/t = \frac{1}{2}$ from Fig. 45. With the new material, a 50-μm-thick plate, and a 25-μm gap, the $\lambda/4$ voltage would be about 15 V, if one assumes that an electrode width comparable to the thickness of the ceramic plate is used.

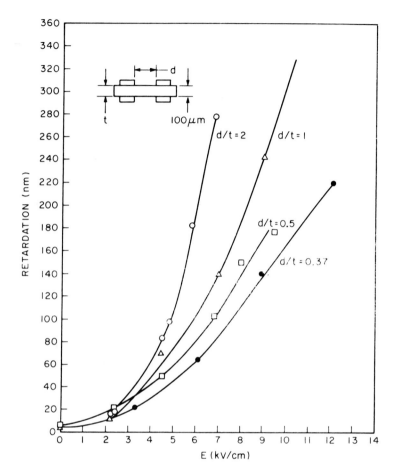

Fig. 45.   Retardation versus applied electric field in a 100-μm-thick plate of PLZT 9/65/35 for several values of the d(gap)/t(thickness) ratio.

A device having a 250-μm separation interdigital array and opaque gold electrodes 75-μm wide was constructed using a polarizer with a built-in λ/4 plate. The contrast ratio measured by a spot meter positioned normal to the ceramic plate was 10:1. The contrast ratio at 40° off the normal was 5:1. The contrast ratio obtained using transparent electrodes with one polarizer and no λ/4 plate is limited by the ratio of gap to electrode area. This effect can be eliminated by using thin opaque electrodes and a λ/4 plastic plate (as mentioned above) with some decrease in viewing angle. The limitation in contrast ratio with transparent electrodes could also be eliminated by using the device between crossed polarizers.

b. Devices with Staggered Arrays. The device structure is shown in Fig. 48. The width of the transparent electrodes is made equal to the electrode spacing (this condition may not be necessary), and the electrode region on one side of the plate overlaps the gap region on the opposite side. The ceramic plate is made thick enough (thickness > gap) in order that the interaction between the top and bottom electrodes does not affect the device performance. Arrangement of the electrodes in this fashion reduces the dead area below the electrodes. Therefore, the ratio of dark areas to light areas is maximized, resulting in an increase of the device contrast relative to the white background. The construction processing is similar to the one described in Section A. This structure required larger voltages than the one described in A, and its utilization will depend on the realization of a material with a large quadratic electrooptic effect. Nevertheless, this structure could find applications in relatively low voltage variable transmission goggles and electronic irises.

c. Engraved Devices. In the devices described above, the applied electric field is not effectively utilized throughout the volume of the ceramic plate. This point is demonstrated by the data shown in Fig. 49. The figure shows retardation versus d/t ratio for several values of the electric field at the plate surface. We observe that when the d/t ratio is decreased, the retardation for a given electric field also decreases. For large electric fields, this effect is more pronounced. This is due to the fact that for small d/t ratios, the field strength at the center of the region bounded by the four electrode edges in a given gap is small, relative to the field on the plate surface. This effect has been analyzed by Baerwald (1969). Therefore, it would be advantageous to have the electrodes engraved through the plate

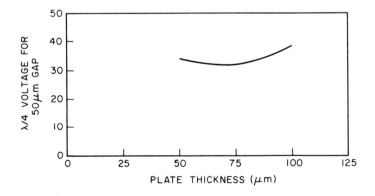

Fig. 46. $\lambda/4$ voltage versus plate thickness for a 50-μm gap.

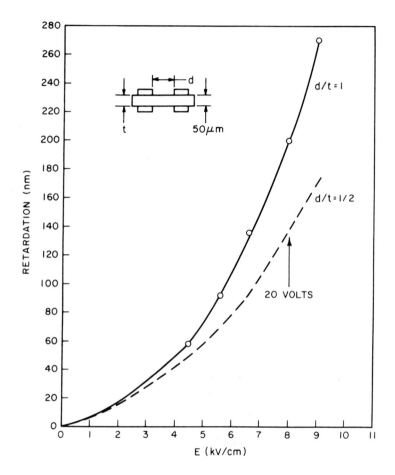

Fig. 47.   Retardation versus applied electric field for a 50-μm-thick sample of PLZT
9/65/35 for $d/t = 1$ and $d/t = 0.3$ (estimated).

thickness. The proposed device structure is shown in Fig. 50. This engraving can be done relatively inexpensively with laser machining using a YAG or an argon laser. Initial experiments using a pulsed YAG laser have been very encouraging. Grooves as deep as 75–100 μm have been machined in thin ceramic plates. Laser machining causes chemical reduction of the ceramic material and the machined region is dark. This condition limits the viewing angle of the device. Fortunately, by oxidizing the ceramic plates (after machining) at about 400°C, the black color disappears and the grooves become almost invisible by blending with the background material. After the plate is laser-machined and heat-treated, indium tin oxide (ITO) transparent electrodes (1-μm-thick) are sputter deposited on the grooved face. Next, the plate is polished lightly to remove the ragged edges and the surface indium tin oxide film, leaving a grid of ITO in the surface. The TiO_2 film and the diffusing mirror are deposited on the grooved side last, and the opposite side of the plate is antireflection-coated. This structure has the advantage that no electrode registration is required. The depth of the grooves and the plate thickness are chosen in such a way that $\lambda/4$ (140 nm) of retardation is obtained with the minimum voltage and maximum viewing angle. Typical numbers are 25-μm-deep, 10-μm-wide grooves on a 100-μm-thick plate; however, more work is required to determine the optimum parameters.

2. *Power Considerations*

The calculations shown in this section apply to the three types of devices described in this paper. It is assumed that each bar is 0.3 $\times$ 0.1 cm and that the gap is 35 μm with electrodes 7 μm wide. Therefore, there are about 80 gaps along each bar. The effective thickness of the plate is assumed to be 50 μm and the relative dielectric constant K of the ceramic is $\sim$5000.

The capacitance of each bar is given approximately by

$$C_{bar} = 80K\epsilon_0 \text{ (area/gap)} \qquad (3)$$

where area equals the thickness times the width of the bar. Therefore, the

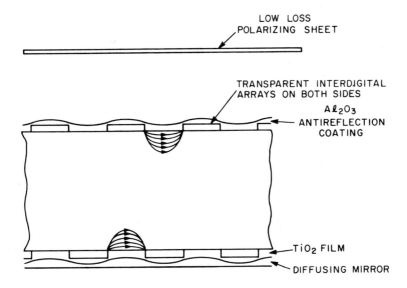

Fig. 48.   Device structure for devices with staggered arrays.

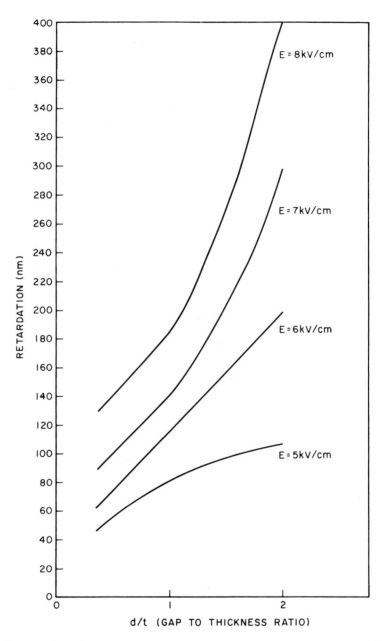

Fig. 49.   Retardation Γ versus d/t ratio for several values of the electric field at the plate surface. 100-μm-thick PLZT 9/65/35.

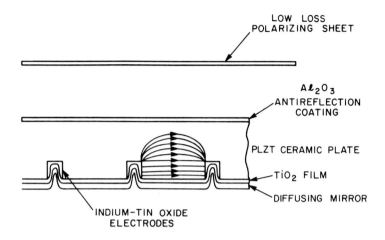

Fig. 50.   Device structure for engraved devices.

charge delivered with 15 V for a seven bar digit is

$$Q = 7 \, C_{bar} V = 15 \times 3.5 \times 10^{-6} = 52 \, \mu C \qquad (4)$$

The instantaneous maximum current for a 0.1-s change is

$$I_{max} = \frac{Q}{t} = \frac{52}{0.1} \, \mu A = 520 \, \mu A \qquad (5)$$

The average power for a 0.1/60 duty cycle (once a minute change) is

$$P = \frac{0.1}{60} \, (0.520 \times 15) \, \mu W \approx 13 \, \mu W \qquad (6)$$

The above analysis assumes that the time constants of the circuit are such that the bars will stay charged for 1 min after the 0.1-s change. This condition is more than satisfied for CMOS circuits with 10 nA leakage in the switch gate. (Experiments have already been done which indicate that the leakage in the ceramic is low enough that a device, once charged, will remain on for a time of the order of days.) We see from this calculation that the power required is of the same order of magnitude as the one needed for scattering-mode liquid crystal alphanumeric devices used in electronic watches.

3. *Performance Limitations*

The data shown above for the devices with interdigital arrays apply to a ceramic plate with the electrode widths comparable to the plate thickness. In order to reduce the insertion loss it is necessary to have the electrode width small relative to the electrodes spacing. (Therefore, the electrode width is made much smaller than the plate thickness.) When this is done, the electrodes no longer behave as indicated by the data shown earlier for a single gap with wide electrodes. The voltages required to switch the ceramic increase somewhat because of the logarithmic dependence of the field inside the gaps. G. A. Coquin (unpublished work) has built some devices using the interdigital array structure with 1-mil electrodes and 2-mil gaps on a 3-mil plate operating with about 40 V. The contrast ratio of the devices was about 5:1 and the insertion loss was relatively high, probably because of the incomplete switching inside the gaps due to the field inhomogeneities produced by the thin electrodes. From his experiments, we infer that the interdigital-array structure will require higher voltages than initially contemplated. Therefore, the engraved device structure may prove to be the most useful of the ones described above for low voltage operation.

Another factor to be considered is the effect of temperature on the device performance. The temperature dependence of the electrooptic effect for the PLZT 9/65/35 composition is shown in Figs. 51 and 52. We observe from Fig. 51 that the electrooptic effect decreases below and above the transition temperature $\sim 15°C$. We also observe from the figure that the remanent birefringence increases below $\sim 15°C$. The increase in remanent birefringence is accompanied by an increase in light scattering which limits the contrast ratio of the device at low temperatures.

Figure 53 shows the normalized applied voltage for $\lambda/4$ retardation at $\lambda = 566$ nm $[V_{\lambda/4}(T)/V_{\lambda/4}(25°C)]$ versus temperature for the same composition used in Fig. 52. We observe from the figure that at 50°C it is necessary to increase the applied voltage by about 30% relative to the 25°C value, in order to obtain $\lambda/4$ retardation at $\lambda = 566$ nm. In an actual device, the effect in contrast ratio with temperature for a fixed voltage might not be as severe as Fig. 53 suggests because of the broad light spectrum of the common viewing sources. The optimum material would be one with a very low transition temperature and a very high electrooptic coefficient at that temperature. The electrooptic coefficient would then decrease above the transition temperature, and a temperature sensitive feedback circuit could be used to increase the applied voltage and compensate for the decrease in electrooptic coefficient. This would then guarantee operation at very low and at relatively high temperatures. More work has to be

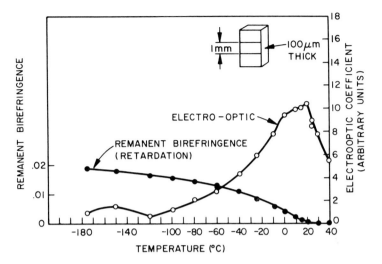

Fig. 51. Temperature dependence of the electrooptic effect for the PLZT 9/65/35 composition. The temperature dependence of the field induced remanent birefringence is also shown. $\lambda = 632.8$ nm, $E_{app} \approx 5$ kV/cm.

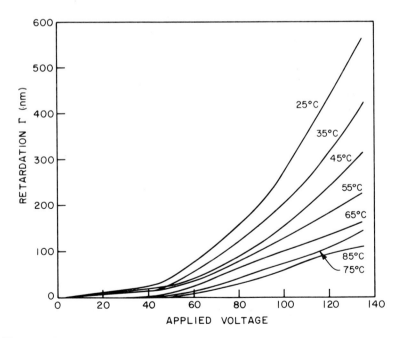

Fig. 52. Temperature dependence of the electrooptic effect above the transition temperature of 15°C in PLZT 9/65/35, 100-μm-thick, 100-μm gap.

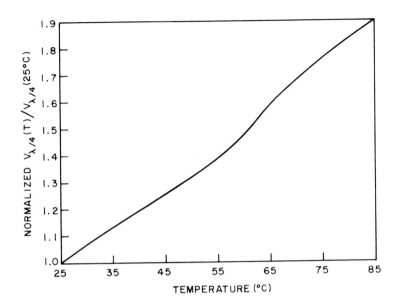

Fig. 53. Normalized applied voltage for $\lambda/4$ retardation at $\lambda = 566$ nm versus temperature for the same composition of Fig. 52.

done to obtain the optimum material; nevertheless, the present materials are capable of operation over a relatively wide temperature range ($-10-+50°C$) with some loss of performance at each end of the range.

4. *Summary*

Several device structures for alphanumeric displays have been proposed using 9/65/35 PLZT ceramics. Among the features of the devices for alphanumeric display applications are the following.

1. They are all solid-state devices.
2. Low operating voltages (<20 V) are possible.
3. Power requirements are small enough (~ 13 μW/digit at 1 change per minute) to enable the device to operate in battery-powered portable apparatus.
4. They are expected to have lifetimes $>10^9$ cycles (more than 30 years when switched once a second).
5. Large viewing angles due to the large index of refraction of the ceramic material (~ 2.5) are permissible.
6. They operate over a relatively large temperature range.

With foreseeable material improvements, it is possible that this technology could be competitive with liquid crystal field effect display devices from the point of view of voltage and power dissipation, with advantages of higher speed, greater temperature range, and probably longer lifetimes. However, the optical transmission of the liquid crystal devices is expected to be better than the ceramic devices because the liquid crystal devices do not use interdigital electrodes. Problems remain in connection with addressing large matrix displays because of the complicated electrode configuration probably needed.

VI. Transparent Conductive and Photoconductive Films for Ceramic Image Storage and Display Devices

A. Organic Photoconductors

1. Polyvinyl Carbazole (PVK)

The photoconductive properties of PVK were discovered by Hoegl et al. (1957). The PVK films, as will become apparent from the discussion below, are not suitable for fast operation (<1 ms); however, because these films are relatively easy to prepare and apply, they are very useful for characterizing ferroelectric-photoconductor devices.

Several authors (Hoegl, 1965; Lardon et al., 1967; Ikeda et al., 1969) have studied the general properties of doped PVK films, and at Bell, PVK films were first used (Lin and Beauchamp, 1970) in a device in conjunction with a thermoplastic holographic storage medium. In this section, we will briefly discuss some relevant properties of the PVK photoconductive films used in the ceramic display devices and describe the technique of coating the devices with the PVK films.

a. Relevant Properties. PVK films with no dopants are very transparent to visible light and sensitive mainly in the UV region of the spectrum (for response curves, see Lardon et al., 1967). With the addition of one of several dopants, the maximum response can be moved into the visible; however, the light absorption in the visible range also increases inevitably. The dopant used in our work is 2-4-7, trinitro-9-fluorenone (TNF) in the ratio of 1 part of TNF to 10 parts of PVK by weight. (Increasing the TNF to a 1:1 ratio improves the sensitivity by a factor of five.)

Figure 54 shows the current–voltage characteristics of a PVK film (4 μm) on gold, with a gold transparent electrode on top, as also shown in Fig. 54 (data taken by Melchior). We observe from the figure (lower

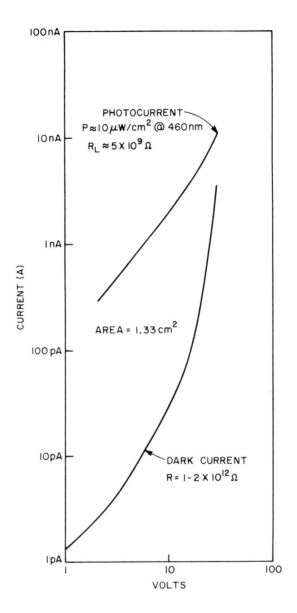

Fig. 54.   Current–voltage characteristics of a 4-μm-thick PVK film. (Data taken by Dr. H. Melchior of Bell Telephone Laboratories.)

curve) that the dark resistance of the cell R_d is about 10^{12} Ω for fields up to about 25 kV/cm. The resistance upon illumination R_L changes to $R_L \approx$ 5 × 10^9 Ω (a 200:1 light to dark ratio). We note from the magnitude of these resistance values that PVK is hardly a photoconductor; it should be more properly called a photoexcited insulator. However, because the resistivity of the ceramic plates is very high ($>10^{14}$ Ω-cm), PVK acts as a reasonable light actuated voltage divider in the ferroelectric-photoconductor devices with ceramic plates typically 50–250 μm. The curves shown in Fig. 54 were taken while increasing the applied field very slowly; if the field is suddenly decreased at any point, hysteresis effects are observed. The light energy necessary to switch the devices using the PVK films about 4-μm thick is about 100 mJ/cm² using a 10:1 PVK–TNF ratio. For a 1:1 PVK–TNF ratio, the necessary energy decreases to about 20 mJ/cm². Some increase in sensitivity was also observed using thinner PVK films (<4 μm).

The following properties of PVK films relevant to ferroelectric photoconductor devices were measured by Melchior: dielectric constant $\epsilon = 2.75$, index of refraction $n \approx 1.8$, quantum efficiency × gain $\approx 10^{-3}$.

b. Preparation and Coating with PVK. The steps followed in preparing PVK and coating with it are given below:

1. Poly-*N*-vinyl carbazole is mixed in a ratio of 10:1* (by weight); 10 parts PVK to 1 part TNF and dissolved in a solution of 50% *p*-dioxane and 50% methylene chloride.

2. Dissolve ~25 g of solid in ~250 ml of solvent.

3. Heat mixture to about 70°C while stirring continuously. (Shield from light while mixing; after mixing, protect PVK solution from light by storing in brown glass bottle.)

4. Viscosity of PVK solution may be adjusted by heating in oven at 70°C to drive off solvent or by adding additional solvent.

5. To coat substrates, dip in a deep vessel half filled with PVK and withdraw slowly using a Fisher-Payne dip coater (Fisher Scientific Company) with cord on fastest pulley. (Glass tank may be obtained from Ealing Corporation, Cambridge, Massachusetts.)

6. When bottom edge of substrate has cleared top surface of PVK liquid, stop dip coater and allow substrate to hang for 1 h in fumes above surface of liquid level. (Resulting layer of PVK should be ~5 μm thick.)

7. Bake PVK layer in oven for ½ h at 70°C at normal pressure.

---

* A mixture with a 1/1 ratio (PVK–TNF) could also be made with the same proportions of the other ingredients.

c. Summary. To summarize this section, we can say that our work at Bell Labs was not directed to characterize the PVK films for device applications. Because of their slow response and high resistivity—which caused image sticking problems due to low charge dissipation—the PVK films were considered only for demonstration of the device capabilities. Due to the high lateral resistivity under electric fields, PVK films are ideal for slow high resolution applications, in which a fast response is not needed.

B. Inorganic Photoconductive Layers

1. *Introduction*

The use of the organic photoconductor PVK was an expedient that permitted testing device concepts. For actual devices, however, the limitations of PVK were too severe and it was necessary to develop deposition techniques for inorganic photoconductive layers. The II–VI compounds are the most widely studied photoconductive inorganic materials to be used in thin film form (Bube, 1967; Cusano, 1967). However, the films have been fabricated predominantly with an electrode geometry such that conduction in the plane of the film is utilized. For a sandwich structure, the necessity of applying the electric field across the film thickness imposes different demands on the photoconductive layer. The most severe of these conditions is that of high electric field. Since the layers are micrometers in thickness, the application of fields in excess of 10^5 V/cm is to be expected in device structures. Initially, high-speed performance with response to microsecond or shorter duration light pulses and conductivity decay times of the order of milliseconds were envisaged for TV rate displays. To make effective use of the addressing light pulses, the density of trapping and recombination centers incorporated in the film should be low. Related to a low density of impurity centers, the photoconductor should have a narrow spectral response in order that projection or ambient light will contribute negligibly to the current through the film. A dark resistivity of 10^{10} Ω-cm is desirable for the dark resistivity in order that contrast in the display device be adequate.

A survey of the various techniques used to deposit CdS led to the choice between vacuum deposition either by evaporation or by sputtering. Deposition by direct evaporation of CdS leads to the growth of a film deficient in S (Cusano, 1967) resulting in low resistivity and low photosensitivity. To compensate for this S deficiency, evaporation techniques may be modified to allow incorporation of more S during film growth. An alternative is that a postdeposition annealing and doping operation (Vecht, 1966) be performed on the film to raise the resistivity and the photosensitivity. The choice was made to sputter deposit CdS (Fraser and Melchior, 1972) because of the superior adherence of sputtered films and because sputtering produces more nearly stoichiometric CdS. In one study by Dresner and Shallcross (1962) where CdS was deposited by evaporation and also by sputtering, it was found that sputtering yielded more nearly stoichiometric films. To reduce the S vacancy concentration still further in the early stages of our display device fabrication, an additional source of S was added to the system. At first, a heated S container was used and this proved to be difficult to control. A mixture of Ar–H_2S was then used as the sputtering gas mixture and this mixture proved satisfactory. The other purpose in trying to maintain the CdS near stoichiometry was to remove the necessity for post deposition doping and annealing. It was believed that such films would be lower in trap density and thus more compatible with fast display systems.

2. CdS Sputter Deposition Process

A diode-sputtering system (Fraser and Melchior, 1972) was utilized, and a conventional vacuum system consisting of a 3-inch oil-diffusion pump charged with Dow Corning 705 fluid and trapped liquid N_2 was used to pump on the chamber. The chamber was pumped to a pressure of 5×10^{-6} torr before the sputtering gas was admitted. Commercial grade Ar and H_2S gases were used without any attempt to purify the Ar.

Commercially obtained sputtering targets were hot-pressed CdS (Materials Research Corporation) or polycrystalline ingots (Semi Elements Corporation) grown by vapor transport, and consistent similar films were obtained irrespective of the target used. No water cooling was used and this restricted the power levels in the discharge to an average of 30 W for an 8.0-cm diameter target or a power density of approximately 0.6 W/cm^2.

The substrates used for photoconductive film evaluation were Corning 7059, borosilicate glass slides, or glass slides upon which transparent electrodes of ITO, $In_{2-x}Sn_xO_{3-y}$ were deposited. All substrates were cleaned prior to being placed in the vacuum chamber. For uncoated glass surfaces, heated chromic acid with rinses in distilled water, acetone, and methyl alcohol were used. For slides with $In_{2-x}Sn_xO_{3-y}$ electrodes, chromic acid at room temperature was used only if deemed necessary, but the other rinses were routinely used. The slides were dried by a flow of dry nitrogen. Initially, some uncoated glass slides were baked in air at 500°C after cleaning, but this did not appear to be necessary since adequate film adherence was obtained without baking. For PLZT substrates, solvent cleaning and an anneal at 500°C in air for 20 min was adequate preparation.

The substrate-to-target distance was approximately 3 cm. For dc sputtering, typical potential differences between the CdS cathode and ground, which acted as the anode, ranged from 1 to 2 kV with current values of approximately 10–30 mA. Generally, deposition rates of 1 μm/h were achieved. Occasional use of rf sputtering yielded similar deposition rates though somewhat lower pressures (20–30 mtorr) and larger target-to-substrate spacings (5 cm) were used. It is estimated that rf power dissipation in the discharge was less than 100 W.

In all CdS depositions, test slides of Corning 7059 glass coated with ITO were placed beside the PLZT substrates. Sandwich-type test structures were then fabricated by vapor depositing In contact dots (10^{-2} cm^2 in area) on top of the CdS layers. The transparent glass and ITO permitted illumination of the CdS layer.

3. *Properties of Sputtered CdS Films*

Examination of the sputtered CdS layers reveals that the films are polycrystalline in the wurtzite phase and possess a fiber texture. The c axes of the grains lie predominantly within an angular range $\pm 8°$ about the surface normal. Varying growth parameters such as power in the plasma, gas pressure, and substrate temperature alter the grain size, usually ~ 1 μm. The polycrystalline nature of the film also results in surface texture which caused light transmission through the film to vary with sputtering conditions.

Adherence of the sputter deposited CdS has been excellent with surface contamination being responsible for any adherence failures. Glass substrates coated with ITO provide the best adherence. Such films have withstood anneals at 500°C, polishing to reduce surface texture, and ultrasonic agitation during solvent cleaning without adherence failure.

The CdS films appear yellow-orange in color. Transmission through a film on 7059 glass is shown in Fig. 55 both in the unpolished and polished conditions. The data have been averaged through the channel spectrum fringes. Note the reduced light scattering in the polished film as the cut-off is approached. Approximately 0.3 μm of film was removed in polishing. To study the effect of substrate temperature, a series of CdS films was deposited with only the substrate temperature varied. Dark and illuminated resistance values were obtained (Fraser and Melchior, 1972).

The CdS was illuminated through the ITO electrode by a microscope light (0.5 W/cm^2 intensity) for the illuminated resistivity (ρ_D) values shown in Fig. 56. All dark-resistivity (ρ_D) values were obtained after the test structure had been in the dark for 10 min with the measuring field maintained. Both illuminated and dark resistivity increase with increasing substrate temperature. However, since the indicated temperature was that of the aluminum block supporting the substrates, the temperature of the growing film would have been higher than the indicated values. It is clear that increasing the substrate temperature did increase both resistivity values, but a preferred range of substrate temperatures exists between 200° and 250°C where high dark resistivity and considerable photosensitivity are both obtained. Another factor that must be considered is that the deposition rate decreases if substrate temperatures in excess of 300°C are used. The deposition rate on a substrate maintained at 320°C was approximately two-thirds of that for a substrate at 220°C. In device fabrication, the CdS films were deposited on PLZT substrates maintained in the preferred temperature range of 200°–250°C. Dark resistivity of CdS on device structures always exceeded 10^7 Ω-cm and usually approached 10^{10} Ω-cm.

To measure the effects of doping, an 8-μm-thick CdS film was deposited on a substrate at approximately 200°C using Ar–6% H_2S as the sputtering gas. One portion of the film sample was prepared for testing by vapor depositing the In dot contacts without treating the film. Another portion was treated prior to deposition of the In contacts. The doping consisted of heating the film sample at 500°C for 30 min in a vitreous silica enclosure to which $CdCl_2$ and CuCl powder had been added. The dc characteristics of the two film samples are shown in Fig. 57 for both dark and illuminated conditions. White light illumination of 0.5 W/cm² intensity was used. Note the overall lower resistivity of the doped film in both dark and illuminated

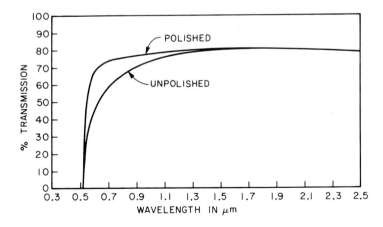

Fig. 55. Transmission characteristic (smoothed) of sputtered CdS on 7059 in the polished and unpolished condition. (After Fraser and Melchior, 1972.)

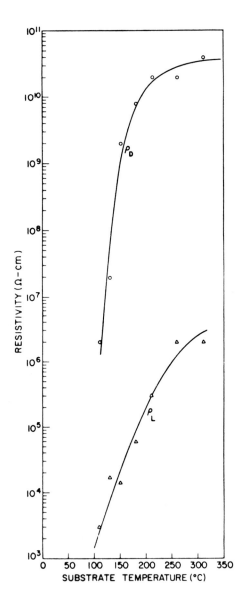

Fig. 56.   Resistivity values in the dark and light for CdS sputtered with H_2S–Ar on heated substrates. (After Fraser and Melchior, 1972.)

conditions. A positive lower recombination center density in the treated film would account for its higher effective gain. Photoconductive gain is usually defined as the number of electrons passed through the film per *absorbed* photon. For convenience, since the quantum efficiency of the process is not known, an effective gain defined as the number of electrons passed through the film per *incident* photon is used. However, when pulse illumination is used, this simple interpretation is subject to question. Pulse illumination of the two films was also performed and their responses are compared in Fig. 58. For both films, the current flowing through a 100-Ω load resistor is shown for the same excitation with 0.4 mW light pulses of 150 μs duration and 40 Hz repetition rate from an argon laser at a wavelength of 5145 Å. In the as-deposited film, the current gain—i.e., the number of electrons collected per incident photon—is of order unity for a bias voltage of 2.5 V. Copper treatment increases the current gain of this film somewhat. As can be seen from the lower trace of Fig. 58, the same current (i.e., unity gain) is reached at a bias voltage of 0.6 V. The current gain of the treated film thus increased by a factor of 4. The risetime of the photocurrent is 50 μs for this excitation intensity and is not altered appreciably

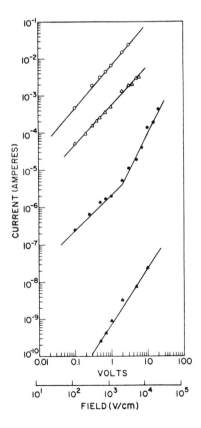

Fig. 57.   Logarithmic plots of *I* versus *V* for a CdS film (∼8-μm thick) as deposited (△, light; ▲, dark) and treated (○, light; ●, dark). (After Fraser and Melchior, 1972.)

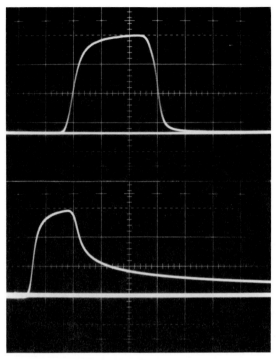

Fig. 58.   Photocurrent in a treated (upper) and an untreated CdS film when pulse illuminated by 514.5-nm Ar laser light. Vertical scale 50 μA/div in both photographs. The horizontal scale is 50 μs/div in upper photograph and 0.1 ms/div in lower photograph. (After Fraser and Melchior, 1972.)

by the copper treatment. The decay time of the copper-treated film, however, becomes much longer. While the duration of the initial part of the decay increases from about 60 μs in the untreated sample to over 100 μs, a long tail extending over more than 1 ms is observed in the copper-treated film.

Thus, while CuCl treatment increases the photoconductive gain slightly, it gives rise to a much longer decay transient and decreases the dark-to-light resistance ratio. Short transients have been obtained in H_2S-treated films that also have high gain.

An example of the pulse response of such a high-gain film is shown in Fig. 59. This film was deposited with an indicated substrate temperature of about 300°C using an Ar–2% H_2S gas mixture. The current flowing in a 100-Ω resistor is shown for excitation with 35-μW light pulses of 150-μs duration at a repetition rate of 20 pulses/s from an argon laser at 5145 Å.

With 2 V applied across the 4.3-μm thickness, the maximum current of 10.5 mA represents a gain of about 750 in this film. The rise time is about 150 μs and the initial decay time is 50 μs. A low-amplitude tail is also evident in the decay, which extends less than 1 ms after pulse turn-off. This film demonstrates the high-gain CdS film with short transients that can be obtained using an Ar–H$_2$S sputtering gas.

4. Devices Incorporating CdS Films

Devices with interdigital electrode arrays were the first image storage structures to utilize sputter deposited CdS films. A direct replacement of the PVK (Fig. 4) by CdS was performed with the result that the image could be erased and rewritten. In addition, lower light energy was required and the writing operation was much faster (milliseconds) than with the organic photoconductor. The strain-biased form of the ferpic was then fabricated using sputter deposited CdS as the photoconductive layer. Replacement of the organic PVK photoconductor yielded devices which were erasable, equivalent in resolution, more economical in their use of writing light power, and faster. A strain-biased device was operated at near TV rates with CdS photoconductive layers (Melchior *et al.*, 1970).

5. Sputter Deposited Cd$_{1-x}$Zn$_x$S

The advantages offered by the alloyed photoconductor Cd$_{1-x}$Zn$_x$S make it superior to CdS in many display devices. For example, the band gap is greater than that of CdS and is adjustable through the compositional parameter, x. This may be interpreted to mean that more spectral band-

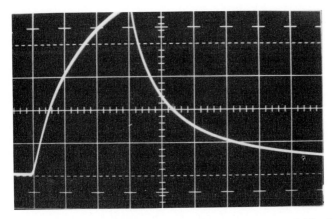

Fig. 59.   Photocurrent in high-gain CdS film. Vertical scale is 2 mA/div, horizontal scale is 50 μs/div. (After Fraser and Melchior, 1972.)

width is available for projection purposes using the alloyed film. The ability to move the band edge with composition also permits moving the photoconductivity maximum near a given laser wavelength. Some other advantages will be described when specific film properties are discussed.

The method used to deposit the $Cd_{1-x}Zn_xS$ films (Fraser and Cook, 1974) was quite similar to that used to deposit the CdS. Because of higher resistivity, the hot-pressed mixtures of CdS and ZnS used as the targets were only sputtered with rf sources. The targets were all water cooled. A fixed mixture of Ar–1% H_2S was used as the sputtering gas. Sputtering conditions had to be determined empirically for each of the targets. It should be stressed that these conditions were such that the photoconductive film deposition process was compatible with the PLZT substrates. As a result, these photoconductive films do not necessarily represent the optimum for any given composition but do represent what was achieved under the constraints imposed by the PLZT.

If bombardment of the PLZT substrate surface during sputter deposition was too severe the PLZT would itself darken. This darkening was due to a loss of Pb and O. The covering film would also darken indicating contamination from the substrate. Most of the short wavelength photoconductive films were deposited without additional heating, even though comparable power was dissipated in the discharge—an indication that sufficient energy was provided by the discharge process to keep the PLZT at an elevated temperature. As a consequence of this greater flux of energy to the substrate, breakdown of the surface of the PLZT was also more probable. Qualitatively, the energy incident on the substrate increased with the Zn content of the target. Though evident even when CdS films were deposited, tensile stress could be severe in those PLZT plates coated with $Cd_{1-x}Zn_xS$ films ($x \approx 0.5$). Stress levels of greater than 2×10^9 dyn/cm² were found in initial work with the short wavelength photoconductive films. The level of this tensile stress increased with the Zn content of the films. By adjusting the sputtering conditions for each target, the discoloration of the photoconductor-PLZT structure and the stress level were kept at acceptable levels. Generally, device structures had less than 10^9 dyn/cm² tensile stress. The slight curvature due to this lower stress could be balanced by placing a second photoconductor on the opposite face of the PLZT. The benefits of using this symmetric configuration of the photoconductor on the PLZT device have already been discussed.

The deposition rates of the $Cd_{1-x}Zn_xS$ films were similar to those of the CdS: 1 μm or more per hour. As with CdS, surface scattering of light could be reduced by the polishing of some $Cd_{1-x}Zn_xS$ films prior to deposition of the outer ITO electrode layers.

It has been possible to make sequences of $Cd_{1-x}Zn_xS$ films of essentially identical characteristics by carefully controlling the sputtering parameters. As an example, a series of films corresponding to $x \approx 0.5$ could be deposited such that the photoresponse peaks all fall within a spectral range of 2 nm. The effective gain of such films would vary by less than a factor of two over the group. Dark resistivity was also reproducible to within a factor of three.

6. Properties of Sputtered $Cd_{1-x}Zn_xS$ Films

The mechanical properties of the $Cd_{1-x}Zn_xS$ films are similar to those of sputtered CdS films. The films adhere well and tolerate mechanical polishing and vigorous cleaning with solvents and ultrasonic agitation. The hardness of the films is greater than that of the CdS layers.

The optical transmission characteristics for two representative films on ITO-coated Corning 7059 glass are shown in Fig. 60. These $Cd_{1-x}Zn_xS$ films correspond to $x \approx 0.5$ and $x \approx 0.2$, respectively, for the short and long wavelength cut-off. The sharp cut-off is characteristic of the sputtered $Cd_{1-x}Zn_xS$ films after polishing. On the glass substrates, the $Cd_{1-x}Zn_xS$ films appear pale yellow in color. However, on PLZT plates, the $Cd_{1-x}Zn_xS$ films contribute slightly more color due to the interaction between film and substrate (see Fig. 31).

Test samples of these same films used for the transmission measurements above were used in spectral sensitivity measurements and the resultant curves are shown in Fig. 61. In both cases, the ITO electrode was maintained negative and fields of about 4×10^4 V/cm were used during the measurements. A reversal of electrode polarity would result in a shift of the wavelength, corresponding to maximum sensitivity of 2.0–4.0 nm in the long wavelength direction, as well as a decrease in maximum current. Note the sharply defined maximum current region in each curve. The effective gain at maximum sensitivity is 0.8 and 30, respectively, for $x \approx 0.5$ and $x \approx 0.2$. The trend of reduced sensitivity with increasing Zn content of the film indicated by these results has been found to be generally true for sputtered $Cd_{1-x}Zn_xS$. This behavior of the photosensitivity appears to be related, primarily, to a reduced electron lifetime, as well as a reduced mobility in those $Cd_{1-x}Zn_xS$ films containing the most Zn (Fraser and Melchior, 1973).

The films used to obtain Figs. 60 and 61 were also tested in the light and dark to measure the I–V characteristics. These data are shown in Figs. 62 and 63. Again, in each case, the ITO electrode was maintained negative. The illuminating wavelength corresponded to the photosensitivity maximum for the film. Note that in both films current saturation is observed

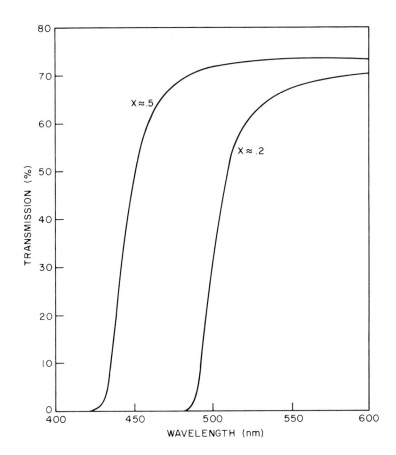

Fig. 60. Transmission of two $Cd_{1-x}Zn_xS$ films with $x \approx 0.2$ and $x \approx 0.5$ corresponding to the longer and shorter wavelength cut-off, respectively.

and is taken to be an indication of coupling between electrons and phonons (Fraser and Melchior, 1973). This current saturation has been characteristic of the $Cd_{1-x}Zn_xS$ films, as well as the CdS films. At fields below saturation, I varies as $V^{1.1}$. In Fig. 63, the point on the illuminated log I versus log V plot at 0.2 V lies off the curve due to photovoltaic effects (Fraser and Melchior, 1973). Prior to making the dark characteristic measurements, the test samples were left in the dark for a period of a few hours under a zero field condition. The I–V data were obtained by increasing the electric field and observing the current after the new field had been applied for $\sim$30 min. This procedure was followed to allow the test film to approach equilibrium between the injected carriers and the populations of the traps for each new field value. At low fields, the dark films behave in an ohmic

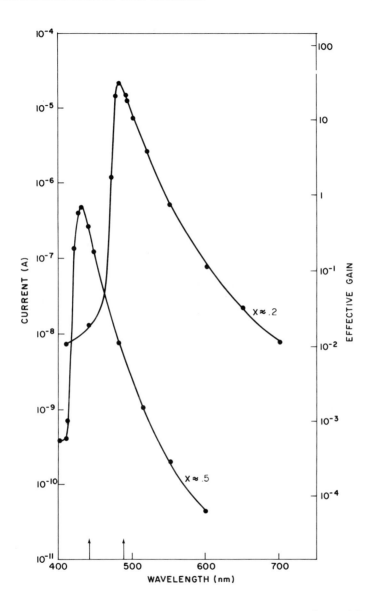

Fig. 61.   Spectral response of $Cd_{1-x}Zn_xS$ films with $x \approx 0.2$ and $x \approx 0.5$ corresponding to the long and the short wavelength photoconductor, respectively. Arrows on the horizontal scale indicate laser lines at 441.6 and 488.0 nm.

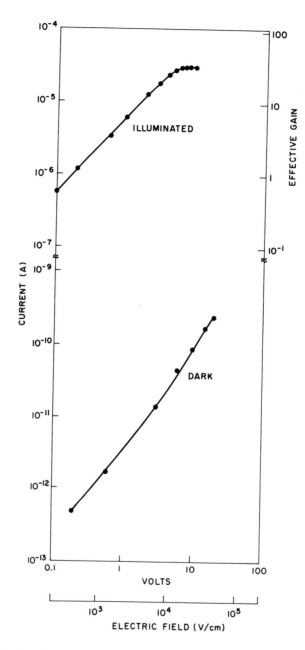

Fig. 62.   Light and dark log I versus log V curves for $Cd_{1-x}Zn_xS$ ($x \approx 0.2$). Film is 4.5 μm thick.

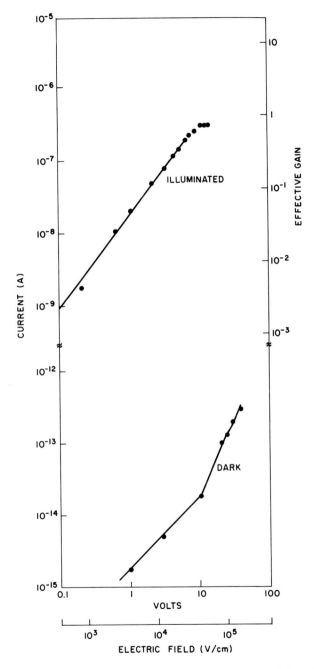

Fig. 63.   Light and dark log I versus log V curves for $Cd_{1-x}Zn_xS$ ($x \approx 0.5$). Film is 2.5 μm thick.

manner. The dark resistivity of the film corresponding to $x \approx 0.2$ was $\sim 2 \times 10^{12}$ Ω-cm, and that of the film with $x \approx 0.5$ was $\sim 2 \times 10^{16}$ Ω-cm. In those applications where information is to be displayed at TV rates, the photoconductive layers must respond to submicrosecond illumination and must also have the dark resistivity recover within milliseconds. This performance has been demonstrated in liquid crystal-photoconductor cells where images have been displayed from a standard TV broadcast (Melchior et al., 1972). A less demanding application is found in the PLZT light-scattering device (Section III). For the operation of this device, a beam of 441.6-nm or 488.0-nm wavelength light is scanned over the photoconductor area, and electrical pulsing at the rate of 500 Hz locally switches the ferroelectric plate.

An example of pulse response is shown in Fig. 64. In this case, a test film sample ($x \approx 0.2$) is pulse illuminated for 1.5 μs by a 488.0-nm wavelength

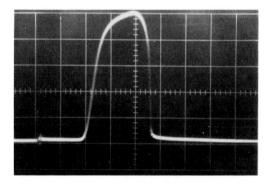

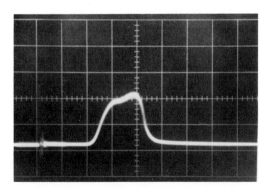

Fig. 64. Current pulse through a $Cd_{1-x}Zn_xS$ film ($x \approx 0.2$). Film thickness is 5 μm. Vertical scale 20 mA/div in upper trace and 10 mA/div in lower trace. Horizontal scale in both photographs, 0.5 μs/div. Laser light of 488.0 nm was used.

laser beam deflected by a PbMoO$_3$ acoustooptic deflector. Two current pulses are shown for illumination levels of 11 and 1.1 mW. At the higher level of illumination the effective gain is 23, and at the lower level the gain is 40. The test was carried out under ordinary laboratory illumination and no discernible change resulted in the current pulse when the room was darkened. Note the sharp rise and fall of the current in the more intensely illuminated case. Fields of slightly greater than 2×10^4 V/cm were imposed across the film. It should be noted that the peak current under intense illumination (laser focused within the 10^{-2} cm^2 electrode area) is approximately 100 mA, and that during such a pulse, sufficient charge could be transferred to switch an area of about 0.03 cm^2 in a PLZT plate. The nonlinear response of the photoconductor is also evident when the two current pulses are compared. When the light intensity is decreased by an order of magnitude, the peak current is only decreased by slightly more than a factor of five due to saturation effects in the photoconductor. However, the rise and fall times at lower level illumination trap filling and emptying are more evident. This has been observed in both CdS and Cd$_{1-x}$Zn$_x$S films showing high photosensitivity. For those films corresponding to $x \approx 0.5$, the effective gain was lower and the rise- and fall-times were approximately 250 ns, essentially the rise and fall times of the light pulse itself. Effective gains of 0.7 have been observed in 1-μs light pulses of 441.6-nm wavelength (Fraser and Melchior, 1973).

An example of the pulse response of a Cd$_{1-x}$Zn$_x$S film with $x \approx 0.5$ is shown in Fig. 65. The illumination was for approximately 20 μs, and corresponds more to the illumination used in a scanned device. A field of 10^4 V/cm was applied across the film. The photocurrent decay was dominated by the shape of the light pulse and is not characteristic of the film. Note the noise evident in the output of the He–Cd laser. The effective gain for the film was about 0.6 (2.8 mW in light pulse). The charge transferred during such a pulse would be sufficient to switch an area of $\sim 5 \times 10^{-3}$ cm^2 in a PLZT plate.

It should be pointed out that the film characteristics described above may be slightly different from the photoconductive properties of similar films on PLZT, since the interaction of the substrate with the growing Cd$_{1-x}$Zn$_x$S film may cause doping. The real test of course is to utilize the film in a device structure.

The properties of the Cd$_{1-x}$Zn$_x$S films have made them more suitable for display applications than the CdS films. A narrow spectral response and reduced sensitivity allow simultaneous READ-WRITE or READ-ERASE operations. The initial high dark resistivity and the rapid recovery of it following laser illumination allow the above operations to be performed at practical scanning rates.

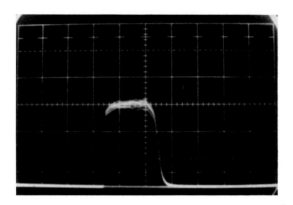

Fig. 65. Current pulse through a $Cd_{1-x}Zn_xS$ film $(x \approx 0.5)$ when illuminated by 441.6-nm laser light. Film thickness of 4 μm was used.

7. *Simultaneous* READ-WRITE *or* READ-ERASE *Operation*

In this section, we will give a quantitative account of the simultaneous reading and writing capabilities obtained when low-repetition-rate, short-duration voltage pulses are applied to a ferroelectric-photoconductor device. The problem will be divided in two parts: (1) an explanation of why cumulative effects (device fogging) do not arise when a series of voltage pulses of duration T and period T' is applied to a fully illuminated device (illuminated from a projection lamp) with no laser addressing, and (2) an explanation of the mechanism for writing by application of a single voltage pulse or a series of pulses to the device while fully illuminated by the projection lamp, and addressed by a focused laser beam.

In order to explain part (1), let us assume that the photoconductor-ferroelectric system, when totally illuminated by a projection lamp, can be represented by the circuit shown in Fig. 66a. The parameters C_{PT} (photoconductor total capacitance), R_{PT} (photoconductor total resistance), C_{FT} (ferroelectric total capacitance), R_{FT} (ferroelectric total resistance) represent the values for the whole (total) plate when continuously illuminated by the projection lamp. In this condition, $C_{PT} \ll C_{FT}$ and $R_{FT} \gg R_{PT}$ (the ferroelectric plate has a very high dielectric constant and a very high resistivity); therefore, the effective time constant of the system is $\tau_1 = R_{PT}C_{FT}$. If a pulse of amplitude V and duration T is applied to the device, the voltage across the ferroelectric plate will be given approximately by

$$V_{FT} = V(1 - e^{-t/\tau_1}) \qquad \text{for} \quad 0 \leq t \leq T \qquad (7)$$

$$V_{FT} = V(1 - e^{-T/\tau_1})e^{-(t-T)/\tau_2} \qquad \text{for} \quad t \geq T \qquad (8)$$

where τ_2 is the time constant of the photoconductor-ferroelectric system when the input is short-circuited. Equation (7) represents the charging of the ferroelectric capacitance through the photoconductor resistance. Equation (8) represents the discharging of the ferroelectric capacitance through

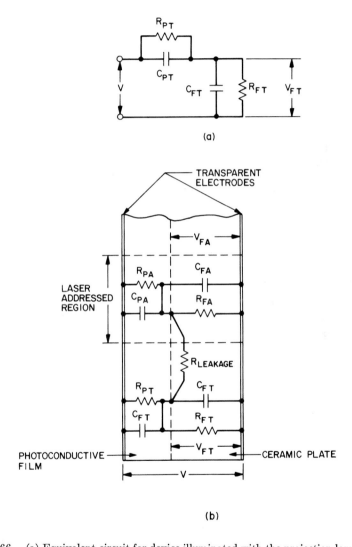

(a)

(b)

Fig. 66.   (a) Equivalent circuit for device illuminated with the projection lamp. The parameters shown correspond to the whole device. (b) Equivalent circuit for device illuminated with the projection lamp and addressed in a small region with a focused laser beam.

the photoconductor resistance when the input is zero (after the end of the input pulse). The back time constant τ_2 in a symmetric device is approximately equal to τ_1; in a single photoconductor device, the photoconductor is addressed to write in the hard (blocking contact) direction and to erase in the easy (ohmic) direction; therefore, it is reasonable to assume that $\tau_2 < \tau_1$. The maximum voltage applied to the ferroelectric is given by Eq. (7) as

$$V_{FT_{max}} = V(1 - e^{-T/\tau_1}) \qquad (9)$$

If $T \ll \tau_1$ then $V_{FT_{max}} \ll V$ and, therefore, less than the coercive voltage (V is normally two times the coercive voltage) is applied to the ferroelectric. In this condition of illumination by the projection light only, very little switching will occur in the ferroelectric. If $\tau_2 \leq \tau_1 \ll T'$ (1/repetition rate), the photoconductor will discharge the ferroelectric capacitance before the next pulse is applied. Therefore, no charge will accumulate in the ferroelectric capacitance to increase the voltage and cause further switching. If $T > \tau_1$ in Eq. (9), then fields larger than the coercive field can be generated across the ferroelectric, and partial switching could take place independent of T', causing fogging of the device. We have observed this behavior experimentally and the optimum pulse width T for 200-V pulses is between 250 and 500 μs. Typically, the pulse width for symmetric devices is about 500 μs and the repetition rate T' is about 500 Hz. The repetition rate could be as high as 1 kHz with no appreciable degradation. If we take $\tau_1 \approx \tau_2 = 500$ μs, then for a 250-μm-thick ceramic plate having an area of 2.5 $\times$ 2.5 cm^2 and dielectric constant $\epsilon = 2000$, the value of R_{PT} turns out to be about 11,000 Ω which is larger than the impedance of the source (pulse generator), typically about 50 Ω and warrants its neglect in the calculations.

The mechanism for writing [part (2)] can be understood from the equivalent circuit shown in Fig. 66b. This drawing shows a laser addressed area of the plate represented by a circuit of the same form but with different components as the one shown in Fig. 66a. This circuit is coupled to the rest of the plate by a resistor, $R_{leakage}$. This $R_{leakage}$ represents the lateral coupling of the addressed spot to the rest of the plate. In first approximation, we can neglect $R_{leakage}$; then the time constant of the addressed area is $\tau_3 \approx R_{PA}C_{FA}$, where R_{PA} and C_{FA} are the resistance and capacitance of the photoconductor and ferroelectric addressed area, respectively.

For our photoconductive films $\rho_{PA}/\rho_{PT} = \tau_3/\tau_1 = R_{PA}C_{FA}/R_{PT}C_{FT} \approx 10^{-3}$ where ρ_{PA} and ρ_{PT} are the resistivities through the thickness of the photoconductor in the addressed area and the rest of the plate, respectively. This indicates that $\tau_3/\tau_1 \approx 10^{-3}$, and for a given pulse width ($\tau_3 < T < \tau_1$) the voltage in the addressed area will rise to a higher value than in the rest of the plate. Therefore, the ferroelectric switching occurs only in the addressed area. In practice, we want to keep T as long as possible to cause maximum switching in the addressed areas with only one pulse and also short enough not to cause excessive fogging as discussed in part (1), above.

C. Sputter Deposited Transparent Electrodes

1. Introduction

For the light valve structures in which PLZT finds application, most devices require transparent conductive electrodes and may require photoconductive layers as well as other layers (reflective or opaque). The basic functions of the transparent electrode are to provide an equipotential surface contact and visible transmission. If the transparent electrode is contiguous to a surface of a photoconductor layer, it will allow suitable charge flow through selected regions of the photoconductor. The selection of these regions can be readily accomplished by means of a scanned laser beam. This combination of conductive and photoconductive films permits selective electrical alteration of the birefringence or the light-scattering properties of the PLZT ceramic. Those regions of the photoconductor not illuminated by the laser do not permit sufficient electric field to be applied across the PLZT to switch it. Another function that the transparent conductor may have to perform is that of a distributed I^2R heating element. Transparent electrodes are used for this application when it is desired to thermally cycle the PLZT between room temperature and some elevated temperature sufficient to depole the ceramic.

2. Transparent Electrodes

In addition to performing the basic functions of transmitting light and making electrical contact, the transparent electrode must have other characteristics suitable for device application. Both chemical and mechanical stability are additional characteristics that must be considered in device applications.

Holland (1966) has extensively reviewed the early work on transparent electrodes. Metal layers deposited by vacuum evaporation from a heated source contribute excessive light loss in a transmission system even when used in thin layers and so are not generally useful. Conductive oxides are an alternative and may be deposited by a variety of techniques, such as pyrolytic chemical reaction, evaporation, or sputtering. All of these methods have been used at Bell Laboratories to obtain transparent conductive films on PLZT.

The disadvantages of the various techniques lead to the choice of sputter deposition of $In_{2-x}Sn_xO_{3-y}$. Some examples of the disadvantages are reduced switching lifetime of the ferroelectric PLZT if pyrolytically deposited films are used (due to the acid environment) and the postdeposition annealing required if oxides are deposited by evaporation. Sputter deposition has yielded ITO films with neither of these disadvantages.

3. *Deposition Process*

A dc sputter deposition technique has been developed which utilizes hot-pressed targets of In_2O_3 plus SnO_2 and a controlled throughput of the sputtering gas, Ar (Fraser and Cook, 1972). A schematic view of the system used to deposit ITO films on PLZT is shown in Fig. 67. The target acts as the cathode and the grounded metal parts of the chamber act as the anode during the discharge. The primary pump was an oil diffusion pump used without liquid N_2 trapping. To minimize backsputtering of oil, the system was roughed through the same restriction used to control the sputtering gas throughput. Argon was allowed to flow through the system during all phases of pumping with the diffusion pump.

The target composition, In_2O_3 plus 9 mol% SnO_2, was decided upon after a series of deposition experiments (Fraser and Cook, 1972). A number of targets of differing composition were used under a fixed set of sputtering

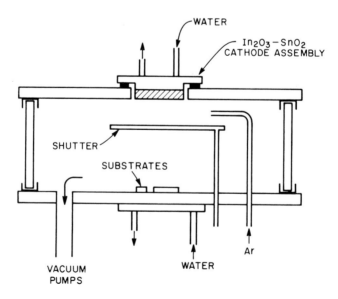

Fig. 67. Schematic view of sputtering station used for deposition of ITO.

conditions and the resultant films were compared on the basis of bulk resistivity values. It was found that a broad minimum in resistivity occurred for those targets whose composition was near In_2O_3–10 mol % SnO_2. Fabrication of the target was somewhat easier for the supplier at low SnO_2 content, and this favored the choice of 9 mol % SnO_2. The PLZT substrates were placed directly on the water-cooled floor of the sputtering chamber to prevent their overheating during the film deposition. Test films were deposited simultaneously on glass substrates contiguous to the work piece.

The sputtering parameters generally used were similar to those for deposition of other ITO films (Fraser and Cook, 1972). The flow rate of Ar through the chamber was maintained at about 0.10 torr liter/sec at a power level corresponding to about 0.8–1.0 W/cm^2 at the target. This low level of power was used so that excessive heating of the PLZT surfaces being coated would be avoided. In the case of ITO deposition on PLZT previously coated with a photoconductor (either CdS or $Cd_{1-x}Zn_xS$), the low power of the plasma reduced the interaction between the layers during deposition.

4. Properties of ITO Films

The properties to be described are those of ITO test films deposited on fused quartz plates while PLZT $Cd_{1-x}Zn_xS$ structures were simultaneously coated. The properties of the ITO are strongly dependent on sputtering parameters and the nature of the reactivity of the substrate. As a general rule, for the low power sputter deposition of ITO—such as used for the PLZT devices—the ITO on fused quartz had sheet resistance values within 20% of the resistance of ITO films deposited simultaneously on PLZT.

In Fig. 68 a plot of transmission versus wavelength is shown, and in Fig. 69 a plot of reflection versus wavelength is shown. Comparison of the two curves shows that the light loss in the visible for the film is mainly caused by reflection. The film was 0.14-μm thick.

The measured sheet resistance of the film was 100 Ω/square. With the film thickness of 0.14 μm, this sheet resistance yields a calculated bulk resistivity of 1.4×10^{-3} Ω-cm. This bulk resistivity was satisfactory for ITO films deposited on PLZT device structures where heating is not a normal function of the ITO layer. For those ITO films required to act as distributed heating elements, sheet resistance values of 50 Ω/square or less were required. When a photoconductive layer is not present on the PLZT, higher sputtering power levels may be used in the ITO deposition process, yielding routine bulk resistivity values in the film of 4×10^{-4} Ω-cm. With more thermally stable substrates, still high power (corresponding to higher substrate temperature) can yield bulk resistivity $<2 \times 10^{-4}$ Ω-cm (Fraser and Cook, 1972).

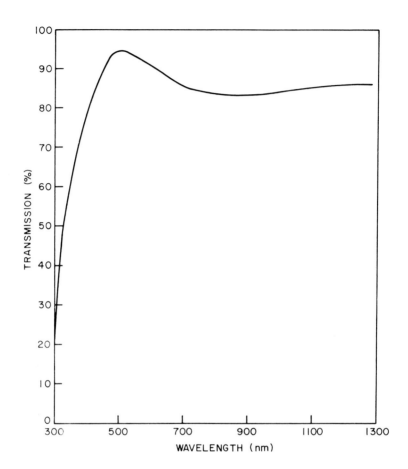

Fig. 68.   Transmission versus wavelength for an ITO electrode film similar to that used over a photoconductive layer. The substrate is vitreous silica.

An example of the optical characteristics of such a heater film deposited on vitreous silica is shown in Figs. 70 and 71. The sheet resistance of this film is 25 Ω/square and its thickness is 0.15 μm. The calculated bulk resistivity of the film is 3.8×10^{-4} Ω-cm. Deposition conditions were such that the power density at the target was about double that used to deposit the film of Figs. 70 and 71.

Some ITO films, usually those thicker than 1 μm, scatter light. When viewed in a scanning electron microscope, these light-scattering films do reveal a surface with grains projecting 10^3 Å above the bulk surface. Thin films such as those used on PLZT devices display negligible scattering.

The primary light loss factor is reflection which, because of $n \geq 2$ for both ITO and PLZT, can only be minimized by selecting the film thickness so as to provide a partial narrow band, antireflection effect.

The ITO films were found to be exceedingly adherent. In the strain-biased form of the device (Maldonado and Meitzler, 1971), the ITO-coated surface of the PLZT was epoxy-bonded to the transparent base material. The adherence of the ITO electrode layer to the PLZT and to the epoxy was crucial to the successful operation of the device. Plates of PLZT coated with ITO have successfully undergone anneals at temperatures of 520°C without film adherence failure.

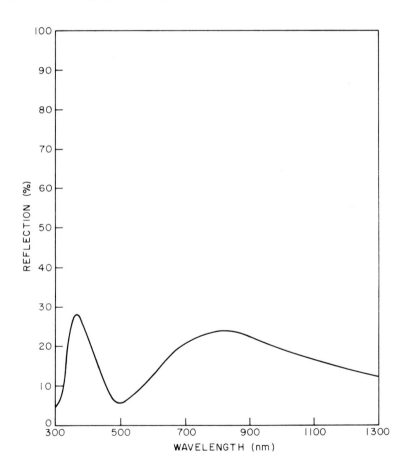

Fig. 69.   Reflection versus wavelength for an ITO electrode film similar to that used over a photoconductive layer. The substrate is vitreous silica.

Definition of ITO electrode patterns has been accomplished in two ways on PLZT plates. The first is a stencil mask technique where a mask of a dielectric material such as vitreous silica has been used to define macroscopic electrodes of millimeter dimensions during the deposition process. The second technique involves photolithographic techniques and selective removal of the ITO by etching. On PLZT photoresist-etch techniques have yielded ITO linewidths of ~10 μm. Oxalic acid has been used as the etchant, since the use of HCl or other strong acids has been found to reduce drastically the switching lifetime (Land and Thacher, 1969) of the PLZT ceramic. This etchant may also be used to selectively etch ITO films de-

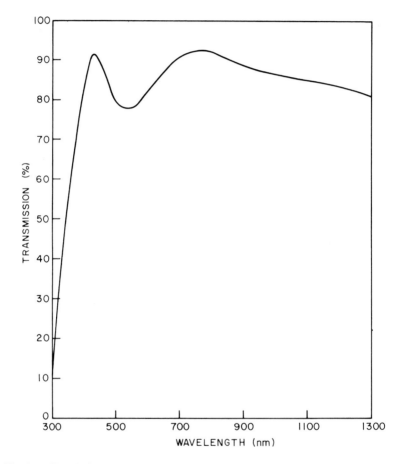

Fig. 70. Transmission versus wavelength for an ITO electrode film similar to that used as a distributed heater. The substrate is vitreous silica.

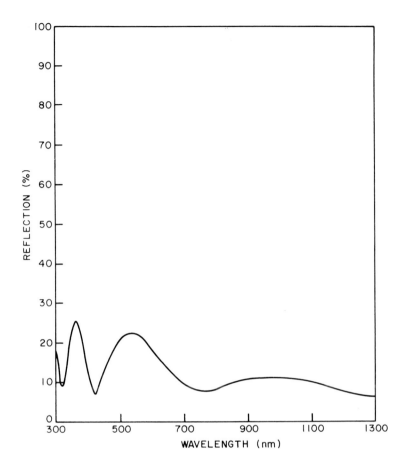

Fig. 71.   Reflection versus wavelength for the ITO film of Fig. 70. The substrate is vitreous silica.

posited on CdS or $Cd_{1-x}Zn_xS$ photoconductive layers without attack of the sulfide layers.

In those devices where voltage pulses are applied to the PLZT-film structure, durable electrical contact to the ITO film is necessary. Relying only on the random point contacts of a metal lead pressed against the ITO results in degradation of the ITO. Severe chemical reduction occurs at the contacts and eventual erosion of the ITO leads to intermittent and open contacts. A suitable contact may be formed by first vapor-plating a gold–chromium tab on the ITO, and then contacting the lead to the gold layer.

VII. Concluding Remarks

This article indicates that a number of display device structures using PLZT ceramics have been proposed and developed. Because of the wide range of useful electrooptic properties obtainable from PLZT ceramics, the structural details of the electrooptic devices can take on considerable variation, depending upon the specific application and the mode of operation of the ceramic. Nevertheless, all of the applications of PLZT ceramics discussed here invoke basically a thin plate of PLZT working in combination with one or more film layers of electrode or photoconductive materials. As with most solid-state devices, the level of performance that can be achieved, and such important ancillary characteristics as lifetime and stability, depend critically upon the control and perfection of the materials used in the devices.

Successful PLZT device development requires sophisticated ceramic polishing and thin-film deposition facilities. Several of the PLZT devices described in this article have been developed to the point that their performance capabilities are useful for practical applications. What is difficult to predict at the time of this writing is whether or not the development of any one of these PLZT devices will progress to the point that it is manufactured for sale in the commercial market place.

In Section III, a slow scan system for a remote blackboard application was specifically described. This system, in the opinion of the authors, represents one case where the advantages of the PLZT ceramic materials result in a device that can compete with any of the other display devices practical for this application, and illustrates the important point that the eventual usefulness of any new display technology is very much dependent upon the requirements of the specific display system in which is going to be used. In the future, there might be other specific display systems which can employ to advantage the unique properties of the PLZT ceramic materials. There is another important point which must be kept in mind in attempting to gauge the eventual usefulness of the PLZT display devices: over the years during which transparent PLZT ceramics have actively been developed for use in electrooptic devices, a truly remarkable improvement in the range of properties and capabilities has been uncovered and the possibility still exists for further development.

In conclusion, the authors are optimistic about prospects for the further commercial development of PLZT ceramic display devices. Since the future cannot be accurately forecasted, the purposes of this paper have been to record the development of basic device ideas, to explore the performance capabilities of the more promising device structures known at present, and to review briefly the fabrication techniques and the performance capabilities of the films essential to display device operation. The authors hope this chapter satisfactorily accomplishes these objectives.

Acknowledgments

The authors wish to thank L. K. Anderson for very valuable comments during proofreading of the manuscript; G. A. Coquin, C. E. Land, M. Melchior, and H. M. O'Bryan, Jr. for very interesting discussions of their work; J. W. Farrell, T. H. Lalonde, W. J. Nowotarski, and C. J. Schmidt for technical assistance; and J. Thomson for the fabrication of the ceramic materials.

References

Baerwald, H. G. (1969). Sandia Labs Res. Rep. SC-RR-69-85.

Bube, R. H. (1967). "Photoconductivity of Solids." Wiley, New York.

Coquin, G. A. (1972). *Opt. Soc. Amer. Meet., San Francisco, Calif., 1972.*

Cusano, D. A. (1967). *In* "Physics and Chemistry of II-VI Compounds" (M. Aven and J. S. Prener, eds.), Chap. 14, pp. 709–766. North-Holland Publ., Amsterdam.

Delissa, A. L., and Seymour, R. J. (1973). *Proc. IEEE Spec. Issue Display Devices* **61**(7), 981.

DiSabato, J., and Fraser, D. B. (1973). Wescon 1973, Session 30, paper No. 5.

Dresner, J., and Shallcross, F. V. (1962). *Solid-State Electron.* **5**, 205.

Fraser, D. B., and Cook, H. D. (1972). *J. Electrochem. Soc.* **119**, 1368.

Fraser, D. B., and Cook, H. D. (1974). *J. Vac. Sci. Technol.* **11**, 56–59.

Fraser, D. B., and Maldonado, J. R. (1970). *J. Appl. Phys.* **41**, 2172.

Fraser, D. B., and Melchior, H. (1972). *J. Appl. Phys.* **43**, 3120.

Fraser, D. B., and Melchior, H. (1974). Unpublished.

Haertling, G. H. (1971). *J. Amer. Ceram. Soc.* **54**, 303.

Haertling, G. H., and Land, C. E. (1971). *J. Amer. Ceram. Soc.* **54**, 1.

Haertling, G. H., and Land, C. E. (1972). *Ferroelectrics* **3**, 269.

Haertling, G. H., and McCambell, C. B. (1972). *Proc. IEEE Lett.* **60**(4), 450.

Hanlet, J. M. N. (1963). U.S. Patent No. 3,083,262.

Harris, J. O., and Cutchen, J. T. (1972). *SID (Soc. Inf. Display) Dig., San Francisco, Calif.* p. 16.

Hoegl, H. (1965). *J. Phys. Chem.* **69**(3), 755.

Hoegl, H., Süs, O., and Neugebauer, W. (1957). Ger. Patent No. 1,068,115.

Holland, L. (1966). "Vacuum Deposition of Thin Films." Chapman & Hall, London.

Ikeda, M., Morimoto, K., Murakami, Y., and Sato, H. (1969). *Jap. J. Appl. Phys.* **8**(6), 759.

Land, C. E. (1967). Sandia Labs. Reprint SC-R-67-1219.

Land, C. E. (1970). *Int. J. Nondestructive Test.* **1**, 315.

Land, C. E., and Smith, W. D. (1973). *Appl. Phys. Lett* **23**, No. 2, 57.

Land, C. E., and Thacher, P. D. (1969). *Proc. IEEE* **57**, 751.

Land, C. E., Thacher, P. D., and Haertling, G. H. (1974). *In* "Applied Solid State Science" (R. Wolfe, ed.), Vol. 4, pp. 137–233. Academic Press, New York.

Lardon, M., Lell-Döller, E., and Weigl, J. (1967). *Mol. Cryst.* **21**, 241.

Lin, L. H., and Beauchamp, H. L. (1970). *Appl. Opt.* **9**, 2088.

McDowell, C. B., and O'Boyle, L. E. (1971). *Proc. IEEE Fall Electron. Conf., Chicago, Ill., 1971*, p. 97.

Maldonado, J. R. (1971). *West. Electron. Show Conv.*, Pap. 31–3.

Maldonado, J. R., and Anderson, L. K. (1971). *IEEE Trans. Electron Devices* **ED-18**(9), 774.

Maldonado, J. R., and Fraser, D. B. (1973). *Proc. IEEE Spec. Issue Display Devices* **61**(7), 975.

Maldonado, J. R., and Meitzler, A. H. (1970). *IEEE Trans. Electron Devices* **ED-17**(2), 148.

Maldonado, J. R., and Meitzler, A. H. (1971). *Proc. IEEE* **59**, 368.

Maldonado, J. R., and Meitzler, A. H. (1972). *Ferroelectrics* **3**, 169; *IEEE Trans. Sonics Ultrason.* **SU-19**, 169.

Maldonado, J. R., and O'Bryan, H. M., Jr. (1973). *Ferroelectrics* **5**, 281.

Marie, G. (1967). *Philips Res. Rep.* **22**(2), 110.

Meitzler, A. H., and Maldonado, J. R. (1971). *Electronics* **44**, 34.

Meitzler, A. H., and O'Bryan, H. M., Jr. (1973). *Proc. IEEE Spec. Issue Display Devices* **61**(7), 959.

Meitzler, A. H., Maldonado, J. R., and Fraser, D. B. (1970). *Bell Syst. Tech. J.* **49**, 953.

Meitzler, A. H., Maldonado, J. R., and Portnoff, M. R. (1971). *IEEE/OSA Conf. Laser Eng. Appl., Washington, D.C.* Pap. 18.7.

Melchior, H., Fraser, D. B., Maldonado, J. R., and Meitzler, A. H. (1970). *IEEE Conf. Display Devices, New York*.

Melchior, H., White, D. L., and Fraser, D. B. (1972). Unpublished.

Pritchard, D. H. (1969). *RCA Rev.* **30**, 567.

Roberts, H. N. (1972). *Appl. Opt.* **11**(2), 397.

Smith, W. D., and Land, C. E. (1972a). *Appl. Phys. Lett.* **20**, 169.

Smith, W. D., and Land, C. E. (1972b). *SID (Soc. Inf. Display) Dig., San Francisco, Calif.* pp. 14–15.

Thacher, P. D., and Land, C. E. (1969). *IEEE Trans. Electron Devices* **ED-17**, 515.

Vecht, A. (1966). *In* "Methods of Activating and Recrystallizing Thin Films of II-VI Compounds in Physics of Thin Films" (G. Hass and R. E. Thun, eds.), pp. 165–210. Academic Press, New York.

Striped Color Encoded Single Tube Color Television Camera Systems

J. J. Brandinger, G. L. Fredendall, and D. H. Pritchard

RCA LABORATORIES
PRINCETON, NEW JERSEY

I. Introduction—Color Encoding Background

The development of a high performance color television camera that uses a single pickup device represents a significant advance in the state of the art and provides a unique building block for a variety of television systems.

In principle, a color television camera must perform the function of a color analyzer, or scanning colorimeter, by generating electrical signal information relating to the spatial distribution of the red, green, and blue color components of the light issuing from the televised scene. In the three tube camera system (Zworykin and Morton, 1954) commonly used in present broadcast practice, the process is direct and straightforward. Red, green, and blue optical filters are interposed in each of the light paths of three separate camera tubes, respectively. The combined spectral sensitivity characteristics of each tube and associated filter have been established by the National Television System Committee (NTSC) and standardized by the Federal Communications Commission (FCC). The instantaneous amplitudes of the output signals are directly related to the intensities of the red, green, and blue primary color components of the individual scanned picture elements. In operation, the electron beam scanning spot of each camera tube must pass over corresponding points in the focused image at essentially the same time, so that acceptable registration occurs between the red, green, and blue images which are eventually reproduced by a color display device. Misregistration of the images occurs if there is relative geometrical distortion of the scanning patterns, differing scan area (raster) sizes, or miscentering of the rasters. In addition, the stability of light input versus signal output characteristics must be maintained in each camera tube. The limitations of registration, relative tracking, and stability have been successfully controlled in modern camera design by sophisticated optical and electrical techniques; but the degree of perfection, however, is often reflected in physical size, weight, complexity, and cost.

By contrast, as will be pointed out, single tube color cameras can provide essentially perfect registration of color pictures with improved stability and simplified operation in a smaller, lighter weight package, and at lower cost than the conventional three tube camera. It is not implied, however, that any specific single tube camera system is suitable as a replacement in all applications. At present, certain trade-offs or compromises in performance (as will be identified later) must be considered. Hence, a single tube camera is potentially of interest for systems in which the main considerations are cost, portability, space, or simplicity of operation. Live or film camera applications include studio and remote TV broadcasting, surveillance and industrial process monitoring, classroom and studio origination in educational television, local origination in cable television (CATV), home electronic photography, and special military and space systems. Although there may be significant differences in the performance–cost requirements for a camera in the consumer, commercial, and military markets, single tube color camera systems have been successfully employed in each of these areas.

Since only one output channel is usually available from a single tube color camera, some form of encoding is required to enable the three color components to be separated in the output. Single tube color camera systems may be classified according to the type of encoding used to transmit color and luminance information contained in the subject. Two principal encoding system classifications have been extensively investigated: (1) time division multiplex, and (2) space division multiplex (i.e., area-sharing of the camera tube target).

A. TIME DIVISION MULTIPLEX SYSTEMS

The well-documented and historically first system in which a single monochrome camera tube was adapted for color use is the field sequential system. A revolving disk, or its equivalent, containing red, green, and blue optical filters alternately interposed in the light path generates a repetitive sequence of electrical signals representing the red, green, and blue color fields on the photosensitive target. The resulting sequential signal is convertible to the format of a standard simultaneous signal by means of a field storage device, such as a magnetic disk. However, this system suffers from a fundamental problem of "color breakup" with motion. That is, due to motion of the subject, successive fields in red, green, and blue are not identical. When superimposed on a viewing screen, the impression is one of misregistered objects in the three color fields.

A field sequential color system for broadcast purposes that utilized a single tube, rotating color filter camera, was proposed by the Columbia Broadcasting System in 1949. System standards were adopted by the FCC, but were rescinded in 1953 in favor of a compatible simultaneous system, tested and proposed by the NTSC (Federal Communications Commission, 1953), that forms the present standards for commercial broadcast television.

The simplicity of the field sequential system has appeal in certain special applications. An outstanding recent application of a field sequential system has been the transmission of color television signals from space probes and, in particular, from the surface of the Moon. This system involved a mechanically rotating disk, field sequential camera operating in space with the appropriate processing apparatus on Earth that translated the signal to a format compatible with color broadcast standards (Drummond, 1969).

B. Space Division (Target Area-Sharing) Multiplex Systems

Color information contained in an image may be encoded in the signal output of a one tube camera system by subdividing the target into small elements that are individually responsive to separate colors. This is usually accomplished by providing repetitive groups of fine color stripes at the target through which the input light is passed. When the target is scanned in a direction perpendicular to the stripes, separate color signals can be obtained as well as signals representative of brightness information.

1. *Tricolor Vidicon*

An early form of space division or target area-sharing is the "Tricolor Vidicon" (Weimer *et al.*, 1960) developed at the RCA Laboratories between 1955 and 1960.

The ability to generate the three simultaneous signals was achieved by means of a vidicon whose backplate electrode was replaced by three interdigitated groups of stripe electrodes, each group having a stripe filter of a particular color in contact with it. When the continuous film photoconductive target is scanned by a single low velocity electron beam, three simultaneous color signals were made available on separate output leads connected to the individual groups of interdigital electrodes. Tubes of this general construction were developed in both 2-inch and 1-inch sizes, but were not carried beyond the laboratory stage.

An enlarged view of the experimental tricolor vidicon target is shown in Fig. 1. Dielectric color filters which transmit red, green, and blue light are built into the target in a repeated sequence of very fine vertical stripes. On top of the filter stripes are semitransparent conducting signal stripes. Adjacent signal stripes are insulated from each other, but those representing the same color are interconnected by means of "bus bars" to a common output terminal for that color. There are approximately 290 stripes for each primary color, making a total of 870 stripes in a 0.6 inch target or about 0.7 mil width for each stripe.

The signal stripes consist of a thin, semitransparent film of metal registered and deposited on the back of the filter stripes ($\sim$ 0.5 mil wide). The bus bars are relatively thick metallic stripes deposited on the glass substrate. Crossover insulation is provided for the signal stripes by the filters themselves which extend over the inner bus bars. By connecting the signal stripes to the bus bars at both the top and bottom of the target, the possibility of open stripes is greatly reduced. A uniform layer of photo-conductive material is deposited over the signal stripes, as shown in the section view of the target in Fig. 2.

The targets described above were formed by the evaporation of the filter and signal stripe materials through precision fine grills (Gray and Weimer, 1959). The filters used most successfully were of the multilayer interference type consisting of alternate layers of high and low index dielectric materials. In the optical design of these filters, it is necessary to take into account the dielectric material losses, the reflectance of the thin metal signal stripe as well as the optical characteristics of the photoconductor. Figure 3 shows a typical set of transmission curves for each of the primary filters including their conducting stripes.

The basic operation of the tricolor vidicon is similar to that of a mono-chrome vidicon. In the absence of light, the electron beam scans the surface of the photoconductor charging it to the cathode gun potential as in a standard vidicon. Light from the scene passing through the color filter stripes causes increased conductivity of the photoconductor in accordance with the intensity of that primary color component. Three substantially independent charge patterns representing the three primary colors in the scene build up on the surface of the photoconductor during the interval

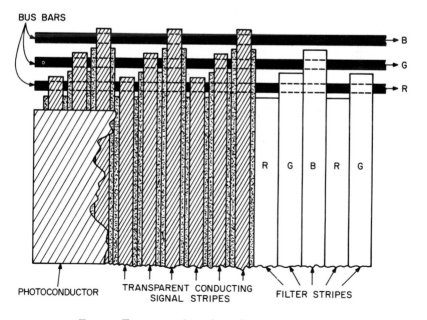

FIG. 1.   Experimental tricolor vidicon target structure.

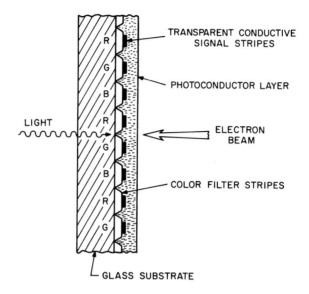

Fig. 2. Sectional view of the tricolor vidicon target.

between scans. The discharge of the three charge patterns by the scanning electron beam results in separate video signal currents simultaneously flowing in each of the three sets of conducting signal stripes corresponding to color signals for each primary color (Borkan, 1960). In practice, large capacitances exist within the target between the three sets of conducting stripes. This results in contaminated color signal components if conventional, high impedance, camera preamplifiers are used. The use of low input impedance amplifiers to avoid contamination generally results in a signal-to-noise penalty. Three to six times more noise compared with that obtained in a conventional monochrome vidicon has been found when satisfactory color signal separation was obtained by using negative feedback to reduce the amplifier input impedance.

A high light loss in the tricolor vidicon occurs due to the light attenuation in the pass band of each tricolor filter and the optical scattering and reflection caused by the combination of filter and signal stripes. Further light loss occurs due to the rejection of the light components outside the pass band of the filters resulting in 25 to 50 times less sensitivity than a monochrome vidicon. However, when compared with a conventional three tube color camera with its inefficient optical system, the tricolor vidicon is only two to three times poorer in sensitivity.

Although of historical and developmental significance, all of the above factors, along with problems of picture streaks due to nonuniformity in target fabrication, excessive lag and optical moiré beats combine to make the usefulness of this approach uncertain.

2. NHK Two Tube Stripe Color Index Camera

A different form of stripe encoding system was developed by the NHK Laboratories in Japan and was successfully used in a two tube camera at the Tokyo Olympic Games in 1964 (Hayashi 1967). This camera contains two vidicons, one for luminance and the other for color. A beam splitter provides the separate light paths required for each tube. For the color tube, spatial encoding is provided by vertical stripe optical filters. These filters are arranged in repeated groups of black (opaque), green, blue, and red (Fig. 4), thus generating a chrominance carrier frequency of 800–900 kHz when the target is scanned. The individual color signals are identified by a particular phase of the color carrier signal. The black stripe located between each color stripe group uniquely establishes a reference time when samples of the green, blue, and red signals are taken, thus assuring independence of sampling from scan nonlinearities. Dark current pulses generated by the black (opaque) index stripes are separated from the camera video signal on the basis of amplitude differences.

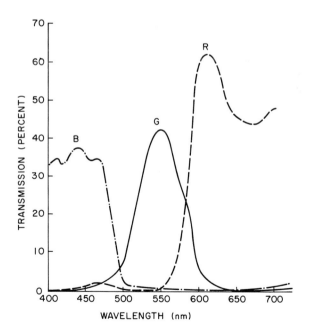

FIG. 3. Typical filter transmission curves for the tricolor vidicon.

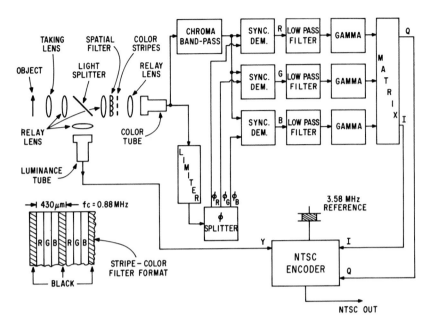

Fig. 4. NHK two tube camera system.

An adverse effect of index stripes is loss in light efficiency and dynamic operating range due to blocking of light by the index stripes. Although sufficient information is available from the striped color tube alone to derive a complete color TV signal, the use of a second tube, registered with the first, allows improved luminance resolution and the use of more easily resolved, wider color stripes. An important advantage of an index encoded color tube is that loss of electron beam resolution, e.g., due to overload in high light areas, results in a less disturbing form of degradation, since a diminished color saturation results rather than a shift in hue. This will be discussed in detail in Section III.

3. *Bivicon Camera Tube Development*

A hybrid form of area-sharing technique similar to the NHK system is incorporated in a novel RCA tube development termed the "Bivicon" (Spalding et al., 1973) which employs two rasters. This device consists of essentially two vidicons in one envelope having two distinct target areas and two independent electron guns. In color applications, it may be operated in a conventional two tube format, whereby one raster is utilized to provide wideband luminance video information, while the other raster is used to provide chrominance signal information by means of any one of several available vertical color stripe encoding techniques.

Three advantages may be gained by combining two tubes within a single envelope (Flory, 1973). (1) The geometric position of the two sets of components of a double tube is fixed within its structure. (2) A single focus coil and deflection yoke are used instead of two sets as in a two tube system. (3) Simplification of the camera optics is possible since the two optical images are positioned close together on the common faceplate containing the two target areas.

A double vidicon considered in the context of a single tube is useful only if it yields a satisfactory degree of stable image registration and if it is economically practical. A recent developmental version of such a tube is in the form of a $1\frac{1}{2}$-inch diameter vidicon with two separate guns and targets. Some novel features include a nonparallel twin-gun structure adapted from the multiple-gun design for color kinescopes, an unusual mesh support design, and a faceplate with a double target structure and embedded signal contact buttons (Spalding et al., 1973). In a system using index encoding for color on one target and luminance for the second, it has been reported that the two rasters retain registry within 1% or less of the picture width.

An important advantage of all two-raster systems is the achievement of wideband luminance signal quality uncontaminated by spurious signals due to spatial color encoding (Goldmark, 1970), as pointed out in greater detail below. A spatial frequency of the color stripes corresponding to an output signal of 2–3 MHz is usually selected such that there is adequate chrominance signal bandwidth and sufficient physical space for the index stripe. The spatial frequency of the index signal is typically one-half of the color signal frequency and is formed in the image by a neutral stripe structure spatially interleaved with the color stripe structure. The color stripes may be designed such that their relative transmissions are equal in a specified "white" light, causing the color carrier to vanish in gray portions of the scene. This condition permits the most efficient use of the dynamic operating range of the system. A chrominance encoding method that provides signal outputs which may be directly translated to the standard NTSC format is usually desirable.

Many versions of two tube color cameras are commercially available. All use the principle of area-sharing of color by means of stripe filters (Sony Corp., 1970; Takemura, 1971; Takemura et al., 1973; Watanabe et al., 1973). Designs differ primarily in whether two or three colors are encoded on the color tube, the width of the stripes, and the efficiency of the optical system. These choices strongly influence the overall cost and performance of the system. Since the principles of stripe encoding are the same regardless of the number of camera tubes, the discussion in the remaining part of the paper will concentrate on the more restrictive conditions imposed by single tube color cameras.

4. *Single Tube Frequency Division Encoding*

Instead of using a single reference carrier and sampling the color signal cyclically at the proper phases of the carrier (as in the case of index tubes described above) the individual color signals may be separated if the color stripes are arranged such that different sets of stripes produce different carrier frequencies representing individual color information. The video output in this case consists of a baseband signal and color subcarriers with attendant modulation sidebands resulting from the target being scanned by the electron beam. In such systems, the color carriers are amplitude modulated in accordance with the intensity of the color information transmitted by the individual color stripes. The resultant output signal is then electronically decoded and the recovered modulation of the spatially generated carriers is matrixed with the low frequency baseband components to form red, green, and blue signals. The signals may then be used to feed a conventional color viewing system.

In practice, a variety of stripe color filter designs and appropriate decoding techniques are possible for frequency-division systems. Representative systems of this type are described in Section II and evaluation data are presented in terms of colorimetry, signal-to-noise ratio, available bandwidth (resolution), decoding circuit complexity, stripe filter type and complexity, light sensitivity, spot focus requirements, intermodulation products, and optics.

II. Frequency Division–Amplitude Modulation Stripe Filter Encoding Systems

A. Dual Carrier System—(Kell/SRI)*

A basic color encoding technique employing multiple carriers was first described by R. D. Kell in 1956 (Kell, 1956; Konig, 1971; Takagi and Nagahara, 1967; Yoneyama, 1972), simplified by A. Macovski (Macovski, 1968a, 1970a, 1972) and was later embodied in commercial live cameras and film chains (Briel, 1970). The essential components of the system are shown in Fig. 5. Light from the subject is passed through a pair of striped optical

---

* Stanford Research Institute.

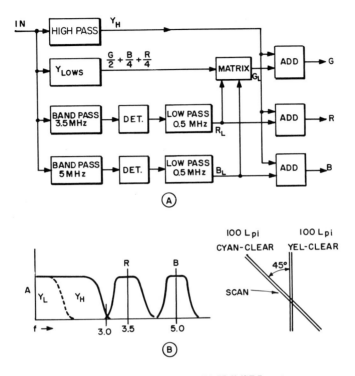

Fig. 5.   (a) Dual carrier and (b) Kell/SRI system.

filters possessing spectral selectivities as illustrated in Fig. 32 (see Appendix I). The encoding action of the filters may be explained qualitatively by visualizing how blue, green, and red components of the televised scene are transmitted by the filters and the resulting charge pattern is scanned off by the camera tube beam. If the first filter is comprised of alternating clear and cyan transmitting stripes, the blue and green light components pass unattenuated to the yellow-clear stripe filter, whereupon the yellow transmitting stripes suppress the blue components and the clear stripes transmit. The amplitude of the modulated square wave along the direction of scanning contains the intensity information of the blue component. Similarly, the yellow-clear stripe color filter acts only on the blue components of the incoming light, passing the red and green light components. When scanning takes place by a beam with a finite aperture, the response is principally a sine wave and a baseband component.

As the vidicon is scanned, two carriers are generated either because the two overlapping gratings, oriented vertically, have different pitches (Kell, 1956) or because one of two equal pitch gratings is oriented vertically while the other is rotated to an angle such as 45° (Macovski, 1972). The two gratings must be designed in such a manner that the resulting signals, as well as the beat between the two signals, will have acceptably low visibility in the luminance pass band. In addition, the grating spatial frequencies must be within the camera resolution limits so that the grating signals may be resolved with an adequate signal-to-noise ratio (Fig. 6).

A convenient method of graphical representation is illustrated in Fig. 7 (Macovski, 1970b). A quadrant circle is drawn in which the radial distance scale is equivalent to the spatial frequency of the grating in cycles per unit distance, and projections on the horizontal axis are the spatial frequency in the direction of scan. If the radius of the quadrant circle represents the spatial frequency of a vertical color stripe whose pitch results in a 5 MHz electrical frequency when scanned horizontally, then the same filter rotated to 45° results in an electrical frequency of 3.5 MHz in the direction of scan as indicated by the projection on the horizontal axis. The length of the line joining the intersections of the quadrant circle represents the spatial frequency of the beat produced between the two gratings.

Spatial frequency in cycles per centimeter projected on the direction of scanning multiplied by the velocity of scan in centimeters per second is the fundamental frequency of the electrical signal resulting from scanning, i.e., $f_x \overset{\circ}{v} = f_{\text{electrical}}$.

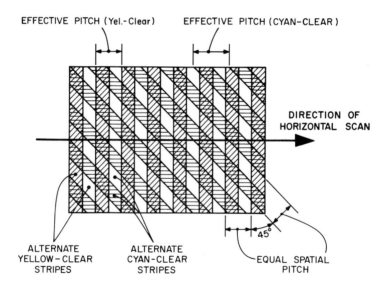

FIG. 6.   Basic color stripe filter relationship.

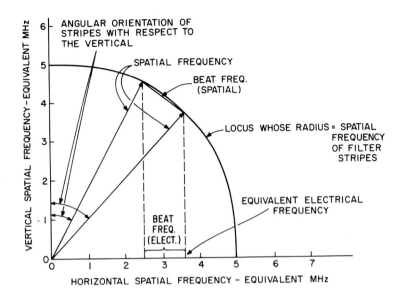

FIG. 7.   Graphical analysis method.

The spectral characteristics of the colored stripes in the striped filters may be specified by relating the camera output signal to the standard NTSC camera sensitivity characteristics. A useful mathematical derivation of the camera output current is developed in Appendix III. An outline of the derivation is as follows: Assume that the televised subject is a uniform monochrome field having a hue designated by the wavelength λ. Let the transmissions of the two striped filters S_1 and S_2 for λ at any point x, y in an image plane in which the filters are placed be $S_1(x, y, \lambda)$ and $S_2(x, y, \lambda)$. Since the sensitivity of the camera target is also a function of λ, a transmission factor $V(\lambda)$ must be included in the expression for the effective light variation $L(x, y, \lambda)$ over the target. $V(\lambda)$ may also include any uniform spatial filter placed at the faceplate of the camera tube. Hence, we write

$$L(x, y, \lambda) = [S_1(x, y, \lambda)] [S_2(x, y, \lambda)] V(\lambda) \qquad (1)$$

When the photosensitive target is exposed to $L(x, y, \lambda)$, a charge is generated according to the relation:

$$Q(x, y, \lambda) = (S_1, S_2, V.)^\gamma \qquad (2)$$

The relation between the beam current $I(\lambda)$ generated by scanning the charge Q is a complex one. In the present analysis, the assumption is made that the signal current is proportional to the total charge covered by a uniform rectangular spot. The result is the following expression, given in terms of system constants, the transmissions $T_c(\lambda)$ and $T_y(\lambda)$ of the colored stripes in the striped filters and the spectral sensitivity of the camera tube, $V(\lambda)$:

$$E(t) = \text{constant } V^\gamma(\lambda)\left[\frac{(1 + T_c{}^\gamma)(1 + T_y{}^\gamma)}{4} + F_1\frac{(1 - T_c{}^\gamma)(1 + T_y{}^\gamma)}{\pi}\right.$$

$$\left. \times \cos(2\pi f_1 t + \psi_1) + F_2\frac{(1 + T_c{}^\gamma)(1 - T_y{}^\gamma)}{\pi}\cos(2\pi f_2 t + \psi_2)\right] \quad (3)$$

where $T_c(\lambda)$ and $T_y(\lambda)$ are the light transmission functions of the color discriminating stripes in the striped filters, and f_1, f_2, x_1, and x_2 are constants in a specific color camera system. These following components are present in $E(t, \lambda)$: (1) a luminance signal proportional to $(1 + T_c{}^\gamma)(1 + T_y{}^\gamma)$, (2) an amplitude modulated carrier $\cos(2\pi f_1 t + x_1)$ generated by one striped filter, and (3) an amplitude modulated carrier $\cos(2\pi f_2 t + x_2)$ generated by the other filter. The luminance signal is separated by low pass filtering and the modulated carriers by bandpass filtering. Envelope detection of the two latter signals recovers information about red and blue picture components. An appropriate combination (matrixing) of the luminance signal with the red and blue signals yields the green signal.

Two serious shortcomings of the Kell/SRI system are the relatively inefficient use of the available frequency spectrum of the pickup device and the colorimetric errors which result from electron beam distortions. The vidicon electron beam spots' shape and size determine whether the color stripes will be resolved, and this in turn controls the amplitude of the red and blue color signal components relative to the green color components contained in the low frequency baseband signal. Thus, as the beam scans the raster area, differences in spot focus and/or astigmatism result in variations of color mixture relationships of red and blue versus green. Mixture colors shift toward the green, and the system "fails-to-green" in the limit when no subcarriers are generated. This limiting condition is encountered in highlight overload or typically in poorly focused corners. Considerable effort, with reasonable success, has gone into the design and development of deflection-focusing systems that result in relatively uniform focus fields. Field uniformity remains as a fundamental problem from both an initial design and an operating point of view for this type of color encoding system.

Other systems have been developed and evaluated that are aimed at minimizing colorimetric variations and improving the bandwidth utilization while still providing acceptable or improved performance in areas such as signal-to-noise ratio, colorimetry, operational simplicity and stability, beats, and resolution (Boyd and D'Aiuto, 1973; Nagahara, 1972; Takagi and Nagahara, 1967; Takemura et al., 1973). Detailed comparisons between the frequency division (Kell/SRI) system and interleaved frequency division systems are reported below.

B. Two Carrier Interleaved Frequency Division—System I

1. System Description

The upper frequency limit usable for color subcarriers in single tube stripe color systems is largely dictated by the available signal-to-noise ratio in the pickup device without reducing the overall bandwidth. For example, the Kell/SRI system utilizes two independent subcarriers for the red and blue color information with one located at about 3.5 MHz and the other at 5 MHz. The baseband upper frequency limit of about 3 MHz results when 500 kHz color channel resolution (1 MHz bandpass for each color) with sufficient guard space has been allowed. It would be highly desirable to relax this upper frequency requirement in the interest of better spot focus uniformity (color uniformity) and improved signal-to-noise performance, while at the same time providing higher resolution in the baseband channel. These seemingly conflicting requirements may be simultaneously met to considerable extent in a system designated as Interleaved Frequency Division—System I (Abrahams, 1954; Brandinger et al., 1974; Pritchard, 1973a).

System I makes more efficient use of the available frequency spectrum by causing the carriers and sideband modulation components representing the red and blue color information to occupy the *same* frequency band. Thus, the upper frequency limit required from the vidicon may be reduced, while an increase is possible in the upper frequency limit of the baseband signal.

The red and blue color signals are caused to occupy the same portion of the frequency spectrum simply by rotating the cyan-clear and yellow-clear stripe filters (equal pitch) such that they are symmetrically disposed on either side of the vertical axis of the scanned raster. Thus, on an instantaneous basis, the carrier frequencies are identical. However, the angular displacement is such that a time delay difference equal to $\pm$ 90° at the subcarrier frequency exists when the signals are compared on a line-to-line basis. This causes the two signals to be frequency interlaced with respect to each other, and the independent color modulation components can be identified and separated from each other by means of a "comb filter" circuit involving a one-horizontal (1-H) delay line (see Appendix II for detailed explanation). The absolute value of the subcarrier frequency is proportional to the pitch of the stripes and to the cosine of the angle of inclination from the vertical. Thus, a relationship exists that allows determination of the angle that results in ± 90° time displacement at the particular carrier frequency involved when one line is compared with the following line. The 180° total displacement between the red and blue signals results in an interleaved energy spectrum and allows for separation by means of a comb filter. Figure 50 (Appendix II) indicates the relationship between spatial frequency, electrical frequency, and angle of inclination required for the line-to-line 180° relationship to exist. Figures 48a and 48b (Appendix II) indicate the vector diagram relationships and the block diagram of the functions performed in a 1-H delay line comb filter circuit.

The required functions for applying this technique to a complete color camera decoding system are shown in Fig. 8a. Figure 8b indicates the frequency spectrum for system I with values appropriate for a system having a luminance bandwidth equivalent to present NTSC color television receiver practices. The value of 4.2 MHz for the color subcarrier frequency, being somewhat lower than the 5 MHz value previously required, places less stringent demands on the beam focus and results in better color uni-

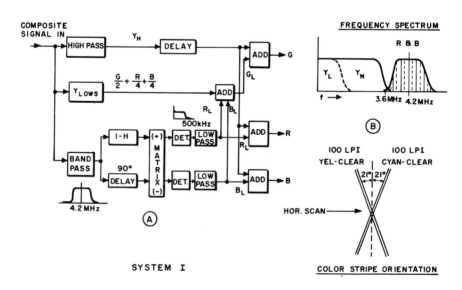

FIG. 8.   Two carrier interleaved frequency division—System I.

formity and improved signal-to-noise ratio, while at the same time allowing the resolution limit of the baseband signal to be increased to at least 3.6 MHz.

As indicated in the block diagram of Fig. 8, only minimal changes from the Kell/SRI system are required in order to realize these performance gains. The same optical stripe color filters are merely reoriented symmetrically at the appropriate angle around the vertical scan axis. The same luminance channel circuits, amplitude detectors, and matrix functions are required. Only one bandpass filter is needed in the selection of the red and blue color information subcarrier. A 1-H delay line comb filter circuit is added for separation of red and blue information prior to envelope detection (see Appendix II).

2. System I Performance Characteristics

As previously stated, system I makes efficient use of the frequency spectrum available from the pickup device. Considerable freedom is available for the choice of specific system parameters, and the system may therefore be effectively tailored to the particular application under consideration. A choice of a color subcarrier frequency of 4.2 MHz along with 500 kHz color bandwidth (1 MHz bandpass) allows for a luminance channel limiting resolution of 3.6 MHz. This particular choice provides an approximate match to the typical response found in commercial color TV receivers.

a. *Signal-to-Noise Ratio.* System I has an inherent advantage over the Kell/SRI system in signal-to-noise (S/N) ratio of at least 3 dB, assuming all other factors to be equal. This advantage occurs because of the interleaving of the two color signals into a common bandpass region, thereby reducing the required color signal bandpass by a factor of two. An additional gain may be realized by reducing the upper frequency requirement from the vidicon. Since the noise spectral density of a video band increases with frequency, an additional S/N increment is gained in system I by the removal of the second subcarrier to a band of reduced noise content. The combination of the two effects results in a measured improvement of 6–8 dB in signal-to-noise ratio. Table I gives actual signal-to-noise measurements at the red, green, and blue decoder outputs prior to gamma reinsertion (see Appendix I).

b. *Colorimetry.* System I and the Kell/SRI system exhibit a similar fault in that variations in the electron beam spot focus leads to varying resolution of the color subcarriers, which in turn causes colorimetric errors that shift toward the green portion of the spectrum. When defocusing

TABLE I

S/N Measurements for Kell/SRI and System I

Color	Kell/SRI	System I
Red	41 dB	47 dB
Green	45 dB	53 dB
Blue	39 dB	41 dB

occurs, the baseband signal containing the green information remains constant while the red and blue information carried by the subcarriers is reduced. However, system I does have improved color uniformity, since the blue and red signal color stripes are equally resolved (same frequency) by the spot and are less subject to differences arising from spot astigmatic problems.

Color matrix adjustments may be made by observing a color bar test scene made of appropriate gelatin filters and causing the red, green, and blue signals to fall within the phase and amplitude tolerances as observed on a vectorscope (Hazeltine Staff, 1956) after re-encoding the signal by means of an NTSC standard encoder. Photographs in Fig. 9a and b indicate typical vectorscope presentation and waveforms for system I set up in this manner.

c. *Beats and Spurious Signals.* Moiré beats occur in the optical systems for the single tube color camera when the high frequency components of the scene content coincide with the pitch of the stripe color filters. These beats must be minimized by a spatial optical filter having a rejection point corresponding to the horizontal component of the pitch of the stripes. Filters of this type are discussed in Appendix I.

"Cross-color" electrical signal beats occur on edges of colored objects that represent the interaction of color modulation sideband components corresponding to high frequency baseband signal components. The amplitude of these beat components is determined by the shape of the baseband modulation-transfer-function characteristic and is usually a compromise between beat visibility and resolution as determined by the system designer.

In the Kell/SRI system, an electrical signal beat of 1.5 MHz may occur between the two subcarriers at 3.5 and 5 MHz and appear as visual beats in large area color mixture portions of the scene. In system I, the red and blue color subcarrier frequencies are the same on an instantaneous basis. However, the signals are separated by comparing each line with the succeeding line in a comb filter (Appendix II). Thus, an interline (7.875 kHz) beat can occur if the filter circuits become sufficiently unbalanced or if the timing of one horizontal scan line with respect to the other varies excessively. Measurements indicate that a timing jitter of 2–3 ns is insignificant, but values above 8–12 ns become objectionable and result in excessive loss of comb filter rejection that appears as improper color rendition. Comb filter circuit balance that results in rejection of the undesired color signal by 30–40 dB is easily achieved with color broadcast quality

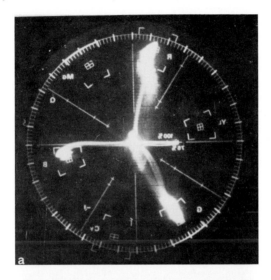

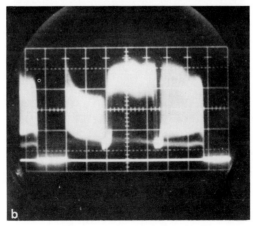

FIG. 9. (a) System I vectorscope presentation and (b) composite signal waveform photo.

sync generator timing circuits. A simple solution to the horizontal timing jitter experienced in some industrial quality sync generators is the use of a "flywheel-driven" camera deflection circuit, which completely eliminates the effects of timing variations on a line-to-line basis. The sync generator source must be locked to a stable source (usually a 3.58 MHz crystal) in order to insure long term stability and matching of the horizontal scanning line period to the 1-H delay line length employed in the comb filter.

C. Three Carrier, Two Interleaved, Frequency Division— System II

1. *System Description*

System I suffers a colorimetry problem if the spot fails to resolve the color stripes uniformly over the scanned area. System II reduces this problem—at some sacrifice in bandwidth utilization efficiency—by introducing an additional color subcarrier at a frequency higher than that of the two interlaced color subcarriers. Thus, all the color information exists as modulation components of the color subcarriers, and the luminance information is carried by the baseband signal. If the subcarriers are not resolved by the spot, the saturation of the scene changes, as opposed to the color hue, and provides a much more acceptable failure mode described as a "fail-to-gray" operation.

Figure 10a is a block diagram of system II, indicating the necessary circuit functions for a complete camera decoder. Figure 10b indicates the required frequency spectrum for system II with the typical values of the frequency parameters.

The red and blue color information is carried by interlaced carriers at 3.6 MHz with equal pitch yellow-clear and cyan-clear stripe filters located symmetrically about the vertical axis at $\pm$ 24°. The red and blue color information is separated by a 1-H delay comb filter, as in system I. A third color signal is generated by the addition of magenta-clear stripes oriented in the vertical direction with a pitch suitable to produce an electrical frequency of 4.92 MHz. This subcarrier provides the green color information and is combined with the red and blue decoded signals in a linear matrix to produce the R–Y, B–Y, and G–Y color difference signals. These signals are then individually combined with the baseband signal to produce the final red, green, and blue output signals.

2. *System II Performance Characteristics*

The most outstanding performance characteristic of system II arises from its fail-to-gray feature. In actual operation, the relative lack of sensi-

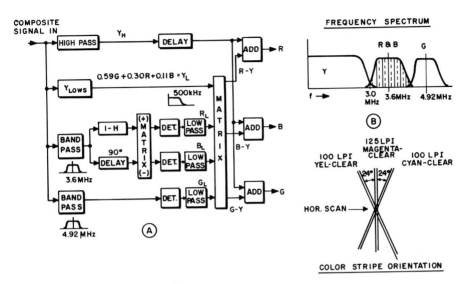

F ɪ ɢ. 10.   Three carrier, two interleaved, frequency division—System II.

tivity to changes in electrical focus is readily apparent since the color
errors occur primarily as changes in saturation as opposed to color hue.

a. *Signal-to-Noise Ratio.* With all elements of the system being
maintained the same except for the addition of the magenta-clear stripe
filter and the appropriate decoder circuit changes, the signal-to-noise ratio
for system II is essentially the same as that of system I. This is true despite the fact that an additional modulated subcarrier has been added to the
system at 4.92 MHz that contains the green color signal information.
Normally, the additional bandwidth at the high end of the spectrum would
degrade the overall signal-to-noise performance. However, three compen-
sating effects result in a signal-to-noise performance comparable with the
two subcarrier systems. First, the interlaced red and blue color carrier
signals are at a lower electrical frequency (3.6 MHz versus 4.2 MHz) and
therefore contribute less noise. Second, the spectral response of the Sb_2S_3
photoconductor peaks in the green portion of the spectrum and more than
offsets the optical loss due to the introduction of the additional stripe color
filter. Third, since the spectral characteristics of the color stripe filters
are designed to produce a baseband luminance signal whose values are in
accordance with the NTSC values (0.59G, 0.30R, 0.11B), the absolute
amount of signal from the green subcarrier required by the matrix to produce the final green output signal is minimized relative to the red and blue
components; thus, the visibility of the noise contained in the green signal

is minimized. Table II gives typical measured values of signal-to-noise ratio experienced in system II (see Appendix I).

The principle of constant luminance (Hazeltine Staff, 1956) may be utilized in the three subcarrier systems to further reduce the visibility of the noise as viewed on a standard color monitor. If the spectral characteristics of the stripe color filters, in combination with the vidicon characteristic, are tailored to produce the NTSC luminance values for the baseband signal (Fredendall, 1973), the decoder matrix circuit must then be adjusted to produce $K_1(R—Y)$, $K_2(G—Y)$, and $K_3(B—Y)$. The values of K_1, K_2, and K_3, when individually added to the Y signal to reproduce the red, green, and blue output signals, are such as to provide constant luminance operation. The reduction in *visible* noise in the viewing monitor has been measured as high as 8 dB. This value can only be realized when the composite signal from the camera has been linearized prior to the decoding and matrix functions.

b. *Colorimetry.* The decoder matrix adjustments may be made by observations on a vectorscope of a color bar scene made of appropriate gelatin filters, and by adjustments of the red, green, and blue matrix parameters such that the signals fall within the NTSC tolerances.

The flat field color uniformity of system II is superior to previous systems due to the fail-to-gray characteristic that translates the partial or complete failure of the spot to uniformly resolve the color stripes into changes in color saturation, instead of color hue. The details of the spectral characteristics are discussed in Appendix III.

c. *Beats and Spurious Signals.* In this system, an electrical beat of 1.32 MHz may occur between the red and blue interlaced subcarriers at 3.6 MHz and the green subcarriers at 4.92 MHz and will appear in large area color mixture parts of the scene. In addition, as in system I, an interline beat (7.85 kHz) occurs if the comb filter becomes sufficiently unbalanced.

The level of the 1.32 MHz beat may be reduced below the limit of visibility if the decoder circuits are linear. However, beat amplitudes of

TABLE II

S/N Measurements for System II

Color	System II
Red	48 dB
Green	51 dB
Blue	46 dB

the order of 3–5% may be effectively eliminated by the use of a simple cancellation technique. Portions of the 4.92 MHz carrier and the 3.6 MHz carrier are selected by two tuned circuits and are mixed in a nonlinear diode circuit to produce a 1.32 MHz beat signal. The beat signal is selected by a tuned circuit and is fed back to an appropriate point in the luminance channel in 180° phase opposition, so as to cancel the undesired residual beat signal. Reduction of the beat signal components between 40 and 50 dB below picture signal white level has been achieved. Figure 11 is a typical circuit for such a cancellation technique.

D. THREE CARRIER, TWO INTERLEAVED, FREQUENCY DIVISION— SYSTEM III

1. *System Description*

Figure 12a is a block diagram of another version of stripe color amplitude modulated systems. Figure 12b indicates the frequency spectrum corresponding to the particular stripe filter orientations required.

As in the previous systems, the red and blue modulated color subcarriers are interlaced on a line-by-line basis at a frequency of 4.2 MHz.

In order to retain the bandwidth utilization efficiency of system I while preserving the fail-to-gray feature of system II, the green color information is carried by a third subcarrier located at a frequency below the red and blue subcarriers. In addition, the green subcarrier (3.2 MHz) is chosen such that the color sidebands overlap and are interlaced with the high frequency detail content of the baseband luminance signal. Thus, the green color information must be separated from the luminance signal by means of a 1-H delay comb filter similar to the unit used for separation of the red and blue signals.

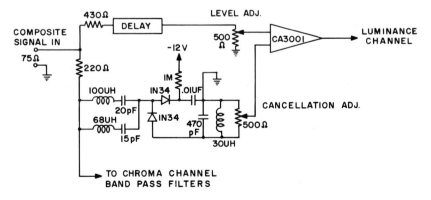

FIG. 11. Beat cancellation circuit diagram.

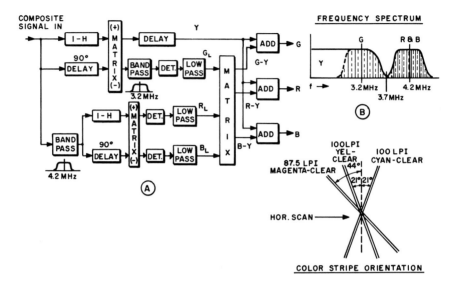

FIG. 12.   Three carrier, two interleaved, frequency division—System III.

The magenta-clear optical filter is oriented at 42° with respect to the vertical with a pitch appropriate for the generation of an electrical frequency of 3.2 MHz in order that the necessary interlaced relationship exist. Either two separate 1-H delay comb filter circuits or a single broadband delay line with the appropriate band pass selection filters may be used for separating the red, blue and green color subcarriers.

2. *System III Performance Characteristics*

System III maintains the highly desirable fail-to-gray characteristic, while making efficient use of the available bandwidth without introducing undue constraints upon the horizontal resolution of the luminance channel. Some loss of resolution does occur, however, along the two approximately 45° directions in the scene, as a result of the comb filter separation of the green color signal from the high frequency luminance components.

a. *Colorimetry.* The large area color field uniformity of this system is subject to variations due to electron beam spot astigmatic effects to a greater degree than previous systems. This result is due to the large angle (42°) required by the magenta-clear stripes. Figure 13 indicates the variation in color subcarrier amplitude versus rotational angle of the color stripes in an actual test setup. This data indicates excessive variations occurring at angles greater than about ± 30°. Subjectively, the fail-to-gray characteristic is an aid in this situation, but the color nonuniformity presents the same problem inherent in the original Kell/SRI system.

E. THREE INTERLEAVED CARRIERS, FREQUENCY DIVISION—SYSTEM IV

1. *System Description*

An interesting and novel approach to the multiple carrier amplitude modulated stripe color system is the location of the subcarriers modulated by the red, blue, and green color information at the *same* frequency, and the use of a 2-H delay line comb filter circuit for separation of the three pieces of color information by comparing *three* successive scanning lines on a line-by-line basis. The relative angles and pitch of the yellow-clear, cyan-clear, and magenta-clear stripe color filters are arranged so that a differential delay equivalent to 120° of phase at the color subcarrier frequency exists between each of the color signals. One example of such a system employs yellow-clear and cyan-clear filters at ± 29° with equal pitch of 100 lines per inch (1 pi). The magenta-clear filter is oriented at 0° (vertical) and has a pitch of 87.5 lpi. Figure 14a is a block diagram of system IV indicating the necessary circuit functions, while Fig. 14b indicates the frequency spectrum.

2. *System IV Performance Characteristics*

System IV, by virtue of having the same subcarrier frequency for all three colors, results in a high order of uniform color characteristics. The

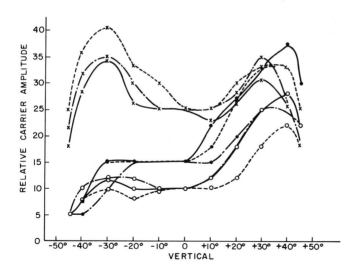

FIG. 13. Carrier response versus angle of rotation of stripes. ●, Left; ×, center; ○, right. - - - - Blue (100 lpi), — – — green (100 lpi), ——— red (100 lpi).

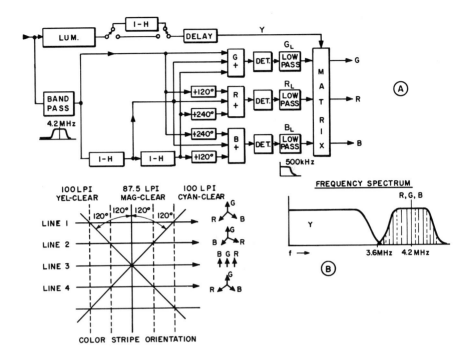

FIG. 14.   Three interleaved carriers, frequency division—System IV.

variations in electron beam spot shape and size affect all three colors in essentially the same way. The system is as efficient as system I in its use of the available bandwidth while at the same time the highly desirable fail-to-gray characteristic of systems II and III is maintained.

On the negative side, the information in system IV is compared on three consecutive scanning lines and the information storage requires the use of two 1-H delay lines in a more complex comb filter circuit. The decoder is more complex and critical to initial design and adjustment. If the comb filter circuit becomes unbalanced, the resulting interline beat of one-third of the line scanning rate is subjectively more visible than the one-half line rate beat in previous systems.

In theory, an additional delay of 1-H should be employed in the luminance channel to provide symmetrical transitions of the vertical detail content; i.e., the luminance transitions should be centered in the three-line period required for a vertical color transition. Subjective tests on typical scenes, however, do not show conclusively that the additional 1-H delay is justified.

All of the interleaved frequency division systems described are capable of acceptable performance when properly implemented, when NTSC broadcast standards are used as the measurement criteria. The fail-to-gray feature of the three color subcarrier systems is distinctly preferable to the colorimetric fail-to-green characteristic of the two color subcarrier systems. Figure 15 is a chart summarizing the various performance characteristics of the representative amplitude modulated subcarrier systems. Each of the systems has individual advantages, as well as limitations that must be considered by the systems design engineer in light of the specific application under consideration. No one system appears to be the panacea for all possible applications. The long list of references describing the variations possible in this general category of stripe color single tube color camera systems attest to the efforts and innovations that have accompanied the development of this color camera technology.

F. Integrated Stripe Filter Vidicon for Frequency Multiplex Systems

The level to which the development of amplitude modulated stripe color single tube camera systems has progressed is exemplified by the "Spectraplex" vidicon developed by RCA Corporation and designated as Vidicon, Type M4445 (RCA, 1972; Serra, 1972), and an integrated filter vidicon developed by Nippon Columbia designated as type C-1102 (Nobutoki et al., 1971).

One approach to encoding systems of this type is to employ a relay optics system in which the actual stripe color filters are mounted external to the vidicon and are imaged at the plane of the photoconductor target by an appropriate lens. All the test results reported above were made in such an optical system (see Appendix I).

A second approach is to construct the vidicon faceplate using fiber optics with the photoconductor, transparent conductive coating on the vidicon beam side, and the color stripe filters cemented optically on the outside of, the fiber optics faceplate. Thus, by virtue of the relatively thin fiber optics, the stripes are translated optically to the photoconductor plane, and the external optics need only consist of the conventional imaging lens. The 4445 vidicon is such a tube in the 1-inch diameter size that contains a set of yellow-clear and cyan-clear color stripes at a pitch of about 530 lpi and oriented at 45° from each other.

In a third approach that has received development effort, the stripe color filters are separately formed on the beam side of a clear vidicon faceplate, and the photoconductor and transparent conducting surface are formed over the top of the filters. Thus, the stripe color filters form an

SYSTEM	KELL/SRI	I	II	III	IV
1. Frequency Spectrum		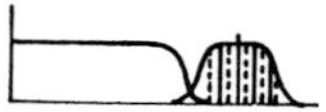			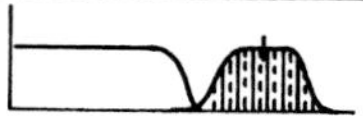
2. Stripe Color Filters	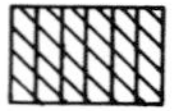 Yellow Cyan	 Yellow Cyan	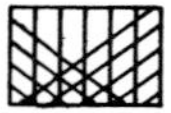 Yellow Cyan Magenta	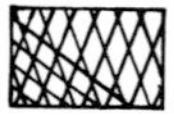 Yellow Cyan Magenta	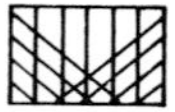 Yellow Cyan Magenta
3. Stripe Orientation	0° @ 5 MHz, 45° @ 3.5 MHz	±21.5° @ 4.2 MHz	±24° @ 3.6 MHz, 0° @ 4.92 MHz	±21.5° @ 4.2 MHz, 44° @ 3.2 MHz	±29° @ 3.85 MHz, 0° @ 3.85 MHz
4. Spatial Frequency	5.0 MHz	4.45 MHz	3.95 MHz – 4.92 MHz	4.1 MHz – 4.2 MHz	4.4 MHz – 3.85 MHz
5. Optical Spatial Filter	3.5 MHz	4.2 MHz	3.6 MHz	4.2 MHz	3.85 MHz
6. Deflection System	Magnetic – GPL 1000	Magnetic – GPL 1000	Magnetic – GPL 1000	Magnetic – GPL 1000	Magnetic – GPL 1000
7. Gamma Compensation	Vidicon Composite Signal Linearizer	← Same	← Same	← Same	← Same
8. Optics	Relay	Relay	Relay	Relay	Relay
9. Vidicon	8507 A	8507 A	8507 A	8507 A	8507 A
10. Comb Filter	None	1-H	1-H	2,1-H(Built), 1,1-H(Poss.)	2-H, Vert. Comp.
11. S/N	R – 40.6 dB, G – 45.2 dB, B – 39.2 dB	R – 47.0 dB, G – 53.0 dB, B – 43.5 dB	R – 47.6 dB, G – 51.1 dB, B – 46.0 dB	R – 45.1 dB, G – 51.1 dB, B – 37.1 dB	R – 47.0 dB, G – 53.0 dB, B – 41.6 dB
12. Luminance Resolution	3.5 MHz	4.2 MHz	3.6 MHz	4.2 MHz	3.85 MHz
13. Chroma Resolution	500 kHz	500 kHz	500 kHz	500 kHz	500 kHz
14. Vert. Color Misregistry	None	1 Line	1 Line	1 Line	2 Lines (Comp.)
15. Large Area Color Uniformity	NTSC ± 20%, ± 10°	NTSC	NTSC	NTSC	NTSC
16. Color Failure Mode	To Green	To Green	To Gray	To Gray	To Gray
17. Comparative Color Uniformity	Poor	Fair	Excellent	Good	Excellent
18. Beats	1.5 MHz, – 36 dB	7.8 kHz, – 40 dB	7.8 kHz, – 40 dB, 1.3 MHz, – 40 dB	7.8 kHz, – 40 dB, 1.0 MHz, – 40 dB	5.25 kHz, – 36 dB
19. Lum./Color Crosstalk (2nd Order)	Magenta Beats	Magenta Beats	Fine Mixture & Interline	Fine Mixture & Interline	Fine Mixture
20. Light Utilization	2/3	2/3	1/2	1/2	1/2

FIG. 15. Summary of characteristics and performance of the amplitude modulation stripe color systems.

integral part of the faceplate target structure inside the vidicon. This type of structure poses some problems in the design of the dichroic filter, comprising the individual stripe color filters, to compensate for the optical termination characteristics of the photoconductor and the transparent conductive coating. The Nippon Columbia vidicon C-1102 is one example of this type.

III. Index Systems—Spatial Frequency Encoding

Another class of single tube stripe color camera systems employing spatial division multiplex may be referred to as "index" systems. In these systems the stripe color filters are usually oriented vertically and allow the red, green, and blue image components to fall on specific areas of the photoconductor for encoding. The output signal spectrum consists of a baseband luminance component and a chrominance component in the form of a single subcarrier which carries color information as amplitude *and* phase modulation. This subcarrier frequency is determined by the pitch of the R, G, and B color group when scanned in the horizontal direction by the electron beam. Low pass filtering isolates the luminance signal and synchronous detection of the color subcarrier by means of a properly phased reference, or index signal, yielding color difference components in a manner similar to NTSC color system demodulation.

Development of this class of systems involves decision and/or innovation in at least two areas. First, the frequency of the index signal relative to the color group frequency must be decided; second, the means of generating a suitable index signal must be determined.

Usually the choice of the R, G, B color group frequency is selected in the upper range of the available spectrum. Typically, a frequency of 4 MHz might be chosen such that adequate luminance resolution may be obtained by a low pass filter that excludes the color subcarrier. The index signal frequency may be chosen to be lower than the color group frequency, exactly equal to the color group frequency, or higher than the color group frequency.

If the index frequency is lower than the color group frequency, a convenient relationship is one-third $(F/3)$. After extraction from the composite signal, the index signal is multiplied by a factor of three and then phase shifted to produce three channels of index signals having symmetrical 120° relationships as required for synchronous demodulation of the color subcarrier. The products of the demodulation are the R–Y, B–Y, and G–Y color difference signals. Two serious problems arise with this approach. First, the index signal frequency is low within the luminance passband and is easily visible as an interfering signal. Second, an effect referred to as "color-phase-pulling" (Thompson, 1960) will occur that results in colorimetric hue errors which vary as a function of the color mixture content of the scene and the relative amplitudes of the primary colors comprising the color mixture. The pulling effect arises from the basic fact that only one index cycle exists for each three color group cycle and the exact time phase relationship between the color group signal phase and the reconstituted index reference phase can be in error during color mixture periods when the composite signal is formed from varying amounts of primary color components.

When the index signal is chosen to be equal to the color group frequency a direct time relationship exists between the color components and the demodulating reference signal. However, a problem now arises as to the means for properly separating the index signal from the color group signal without seriously compromising other performance characteristics such as dynamic range or sensitivity. In this case, the design burden is placed upon the development of a means for generating the index signal.

Two problems arise if a frequency higher than the color group frequency is chosen for the index signal. First, the ability to resolve a high frequency index signal with suitable signal-to-noise ratio becomes increasingly difficult. Second, more than one cycle of the index signal exists for each cycle of the color group frequency, and an ambiguity exists between the index signal and the chrominance subcarrier that must be resolved in some manner on a line-by-line basis

One solution to these problems is the use of two separate pickup devices (Sony Corp., 1970). One tube provides the luminance signal uncontaminated by a color group or index signal, while the second tube performs the functions of generating the stripe color chrominance signal and the appropriate index signal, each being located at a convenient frequency within the pass band of the chrominance tube output. Several camera versions have been developed around this approach including the NHK system and the Bivicon tube previously discussed (Sections I, B, 2 and 3). Numerous other commercially available two tube systems have been reported in the literature but are outside the scope of the present discussion.

A. THE TRINICON INDEX STRIPE COLOR SYSTEM

One example of a stripe color, single tube, index system is the introduction of the "Trinicon" by the Sony Corporation of Japan in 1971 (Kubota and Kurokawa, 1971; Kurokawa and Kubota, 1972; see also Kubota, 1972). The Trinicon tube and the camera system developed around it provide a unique and interesting solution to the basic problems relating to the index approach to a single tube color camera.

1. *Basic Operation*

The basic design concept utilizes a single carrier frequency for the color encoding as opposed to multiple carriers employed in other approaches, and provides a decoding reference (index) signal whose frequency is exactly equal to the fundamental frequency of the color group signal.

The camera utilizes a vidicon type tube, having a filter in the form of vertical triads of striped color selective elements. The target structure, shown in Fig. 16, consists of five principal elements; faceplate; the R, G, and B stripe color filters; an isolating layer; a transparent signal electrode that is formed into interdigitated stripes used to produce the index signal; and the photoconductor layer. A pair of transparent conducting stripes correspond to and are aligned with each triad of color stripes. Hence, the frequency of the index signal and the color subcarrier signal is identical.

The generation of the index signal by the striped transparent conductors is enhanced by the application of small offset voltages of opposite polarity to each of the conductors in a pair. This necessitates two signal output leads to be made available at the target ring. The separation of the index signal from the color group signal is accomplished by switching the polarity of the offset voltages at the horizontal line rate (15 kHz) and using a 1-H delay comb filter technique as the means of identification (see Fig. 17). The polarity reversal on a line-to-line rate of the index signal yields the appropriate 180° relationship from one line to the next. The color carrier is thereby recovered by addition of the information stored in the 1-H

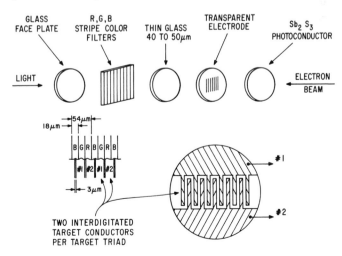

FIG. 16.   Trinicon target structure.

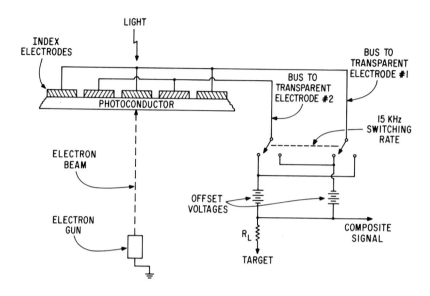

FIG. 17.   Trinicon electronic indexing.

memory with the information in the direct path, while the index signal is extracted by subtraction (inversion followed by addition) of the delayed and undelayed signals. Figure 18 is a block diagram of the complete camera system.

The composite output signal may be mathematically expressed as follows:

$$E_D = E_R(t) \left[a_0 + \sum_1^\infty a_n \cos(n\omega t + \theta_n) \right]$$

$$+ E_G(t) \left[a_0 + \sum_1^\infty a_n \cos\left(n\omega t + \frac{2\pi}{3} + \theta_n\right) \right] \qquad (4)$$

$$+ E_B(t) \left[a_0 + \sum_1^\infty a_n \cos\left(n\omega t - \frac{2\pi}{3} + \theta_n\right) \right]$$

The first three terms, $E_R(t)$, $E_G(t)$ and $E_B(t)$ are the red, green, and blue components of the signal corresponding to the output of the image through each color filter. The remainder of the expression indicates the phase time encoding of the particular color filter stripe. The low frequency components form the luminance signal and may be obtained by a suitable low pass filter. A bandpass filter, centered around the color subcarrier, extracts the chrominance signal that may be shown to be equivalent to a quadrature balanced modulation of two color difference signals. By means of properly related synchronous detection, using the processed index signal as a reference, color difference signal components may be obtained. By proper design and control of both the width and spectral response of the individual stripe color filters, the output signal luminance and chrominance composition may be made such that conversion to standard NTSC values is possible. The Sony Trinicon is reported to employ red, green, and blue transmission stripe color filters. However, it is possible to use the conventional yellow, cyan, and magenta complementary colors to obtain a light sensitivity advantage.

2. Performance Characteristics

The reported specifications and performance characteristics for the 1-inch version of the Trinicon includes a luminance channel bandwidth of 3.6 MHz with the color group frequency located at 4.5 MHz. The color signal bandwidth of 800 kHz (1.6 MHz bandpass) provides a signal-to-noise ratio of about 41 dB with a luminance channel S/N ratio of 52 dB. The sensitivity is said to be ~300 nA of signal current at a scene illumination of 1000 lux (100 fc) with an $F/4$ optical aperture (Kubota and Kurokawa, 1971).

There are no intermodulation beats produced since the color group frequency and the index frequency are identical. Since the luminance

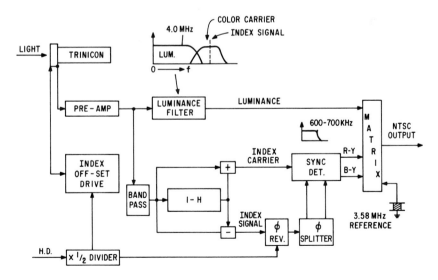

FIG. 18.   Trinicon camera system block diagram.

channel and the chrominance signal components do not overlap, there are no beats on edges resulting from "cross-color" effects.

The index approach to the single tube stripe color camera essentially places the burden of system implementation on the fabrication of the pickup device while employing conventional processing circuitry. First, the individual color stripes of the color groups must be closely controlled in both spectral characteristics and relative widths in order to insure that the color subcarrier balances to zero during a gray portion of the scene. Some correction may be made for small error by means of the color decoding matrix adjustments, assuming the error to be constant and uniform over the scanned area. Second, in the process of fabricating the various stages of the faceplate structure, the registration in the horizontal direction and the relative angular alignment in the vertical direction between the color group stripe filters and the index signal generating stripes must be of a high order. Registration errors are interpreted by the decoding circuitry as color phase errors that manifest itself as hue shifts in the reproduced picture. If $\pm 5°$ is used as a tolerance criteria, as per NTSC practices, requirements for mechanical registration becomes about 1–2 μm. The parallelism of the color stripes and the index stripes must be maintained to within about 0.18 mrad or about 0.01°. Neither of these fabrication problems is outside of today's state-of-the-art technology.

B. Additional Index System Considerations

The following is a brief discussion of the mathematical model for index sampling systems around which the design of appropriate filter spectral characteristics may be developed. If it is assumed that the index signal and the color group frequencies have been either derived, or translated to be equal to each other, then three indexing signals may be formed therefrom to be used as pulse sampling signals for the eventual isolation of the three color video signals designated as A_1, A_2, and A_3.

1. *Pulse Sampling*

The spectral characteristics of the filter stripes may be developed on the premise that amplitudes of the output signals R, G, and B of the matrix circuitry will be a measure of the camera sensitivity curves as unit monochrome light input is varied in wavelength over the desired visible range. Gamma compensation may be assumed and no overall trimming filters need be included as is required by the colorimetry of the amplitude modulation subcarrier systems.

If a highly idealized condition is shown (Fig. 19), a uniform color field $L(\lambda)$ results in an output signal of substantially rectangular waveform.

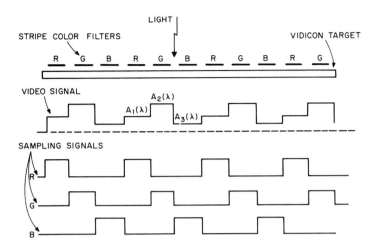

FIG. 19. Idealized camera signal and sampling waveforms for the index stripe systems.

Three rectangular sampling signals, derived from the indexing signal, isolate the blue, green, and red information which is measured by the amplitudes of the sampling signals after multiplication by the respective color amplitudes. If the spectral characteristics of the filter stripes are $B(\lambda)/V(\lambda)$, $G(\lambda)/V(\lambda)$, and $R(\lambda)/V(\lambda)$, where $B(\lambda)$, $G(\lambda)$ and $R(\lambda)$ are the camera sensitivity characteristics and $V(\lambda)$ is the vidicon spectral characteristic, the color $L(\lambda)$ is reproducible on a monitor having the spectral characteristics from which the sensitivity curves were derived, provided that the sensitivity curves do not possess negative lobes. Normalized stripe spectral characteristics are shown in Fig. 20 corresponding to the positive portions of the standard NTSC sensitivities. Camera sensitivity curves for typical rare earth phosphors are shown in Fig. 21.

Since B, G, and R signals are available directly in wideband form from pulse sampling, this version of the indexing vidicon system is analogous to the familiar three vidicon camera. However, the high frequency response of the vidicon is not sufficient at three times the color group frequency to permit sharp pulse sampling. Equivalent operation is possible with sine wave sampling at color group frequency.

2. Sine Wave Sampling

Sine wave sampling of the camera signal is performed in three product or synchronous detection circuits supplied with different phases of a carrier signal of the same frequency as the color group frequency; namely,

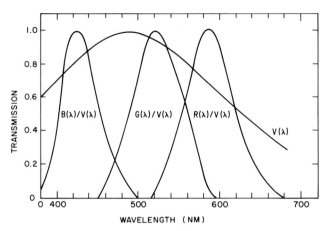

Fɪɢ. 20. Camera sensitivity curves compensated for the vidicon spectral character-istics for the NTSC standard phosphors.

$\cos(x + \theta)$, $\cos(x + \theta + 2\omega/3)$, and $\cos(x + \theta + 4\omega/3)$. For unit incident monochrome light the camera signal shown in Fig. 18 has the form:

$$S(x, \lambda) = \frac{1}{3}\sum_{1,2,3} A_N(\lambda) + \frac{\sqrt{3}}{2\pi}[A_1(\lambda) - 2A_2(\lambda) + A_3(\lambda)]\cos x + \cdots$$

$$+ \frac{3}{2\pi}[A_1(\lambda) - A_2(\lambda)]\sin x + \cdots \tag{5}$$

Only the fundamental signal component is applied to the demodulators. The individual output signals of the demodulators are

$$S_1(\lambda) = A_1(\lambda)\cos(\theta - 60°) - A_2(\lambda)\cos\theta + A_3(\lambda)\cos(\theta + 60°)$$
$$S_2(\lambda) = A_1(\lambda)\cos(\theta + 60°) + A_2(\lambda)\cos(\theta - 60°) - A_3(\lambda)\cos\theta$$
$$S_3(\lambda) = -A_1(\lambda)\cos\theta + A_2(\lambda)\cos(\theta + 60°) + A_3(\lambda)\cos(\theta - 60°) \tag{6}$$

The angle θ is set equal to 60° yielding

$$S_1(\lambda) = A_1(\lambda) - \frac{A_2(\lambda)}{2} - \frac{A_3(\lambda)}{2}$$

$$S_2(\lambda) = -\frac{A_1(\lambda)}{2} + A_2(\lambda) - \frac{A_3(\lambda)}{2}$$

$$S_3(\lambda) = -\frac{A_1(\lambda)}{2} - \frac{A_2(\lambda)}{2} + A_3(\lambda) \tag{7}$$

The luminance signal given in (5) by

$$S_0(\lambda) = \tfrac{1}{3}[A_1(\lambda) + A_2(\lambda) + A_3(\lambda)] \tag{8}$$

Then multiply by $\tfrac{3}{2}$ and add to each member of Eq. (3) to yield $A_1(\lambda)$, $A_2(\lambda)$, and $A_3(\lambda)$.

If adherence to NTSC signal formulation is desired, the luminance signal may be reconstituted from the B, G, and R signals at the matrix output in the form

$$S_0(\lambda) = 0.11B^{1/\gamma} + 0.30R^{1/\gamma} + 0.59G^{1/\gamma} \tag{9}$$

The constant luminance principle may be applied earlier in the system by adjustment of the transmission of the blue, green, and red stripes such that

$$B(\lambda) = \frac{0.11 S_{\mathrm{b}}(\lambda)}{V(\lambda)} K_{\mathrm{M}}$$

$$G(\lambda) = \frac{0.59 S_{\mathrm{g}}(\lambda)}{V(\lambda)} K_{\mathrm{M}} \tag{10}$$

$$R(\lambda) = \frac{0.30 S_{\mathrm{r}}(\lambda)}{V(\lambda)} K_{\mathrm{M}}$$

$B(\lambda)$, $G(\lambda)$, and $R(\lambda)$ are the spectral characteristics of the stripes; $S_{\mathrm{b}}(\lambda)$, $S_{\mathrm{g}}(\lambda)$, and $S_{\mathrm{r}}(\lambda)$ are the camera sensitivities; $V(\lambda)$ is the vidicon spectral characteristic; and K_{M} is a normalizing factor which is adjusted to

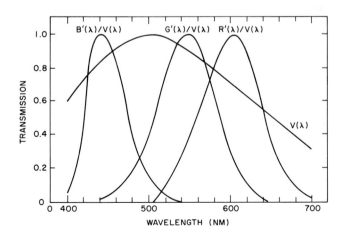

FIG. 21. Camera sensitivity curves for the rare earth phosphors.

attain maximum transmission for $G(\lambda)$ at a given λ. It is evident that the resulting reduction in the visual effect of noise at the display device due to constant luminance must be balanced against the possibility of increased noise in the blue and red channels due to the low multiplying factors in Eq. (10).

An increase in vidicon response for a given light input is achieved when the spectral transmission of the color stripes is complementary to the primary characteristics $B(\lambda)$, $G(\lambda)$, and $R(\lambda)$. These characteristics are equal to the sums of pairs of the primary characteristics in Figs. 20 and 21; that is,

$$\text{(Yellow)} \quad Y(\lambda) = G(\lambda) + R(\lambda)$$

$$\text{(Magenta)} \quad M(\lambda) = R(\lambda) + B(\lambda)$$

$$\text{(Cyan)} \quad C(\lambda) = B(\lambda) + G(\lambda) \tag{11}$$

C. High Frequency Index Signal Systems

If the color group frequency is represented by F, typical index frequency relationships may be $F/3$ in one case, and high frequency fractional relationships such as $\frac{3}{2}F$ or $\frac{5}{2}F$ in other cases. As has already been stated, the $F/3$ situation has two basic problems of color pulling and visibility of the index signal within the luminance band. Although an approach of this nature is ideally suited for two tube versions of index systems, the $\frac{3}{2}F$ or $\frac{5}{2}F$ are generally more suitable for single tube techniques (other than the case of color group frequency and index frequency being equal as previously discussed).

One of the reasons for choosing an index signal at a higher pitch or frequency than that of the color group frequency, depending upon the resolution capability of the vidicon, is based on the need to minimize color pulling effects due to the color modulation of the index signal. Trade-off studies for other indexing systems have indicated that errors are minimized by using specific fractional frequency ratios. As one example, an index stripe frequency of $\frac{3}{2}$ color group frequency will produce a shift of less than $7°$ in the decoding process almost independent of electron beam spot size (Graham *et al.*, 1961). Other ratios like $\frac{4}{3}$ and $\frac{5}{4}$ are possible, but phase shifts due to color pulling are unreported and likely to be small.

The index signal-to-noise ratio is set by the vidicon response and bandwidth if the signal is generated by optical means in the target structure. The narrower the output signal channel bandwidth, the better the signal-to-noise ratio but the longer the integration time constant and the attendant increase in the degree of required scan linearity. For example, for a bandwidth of 0.6 MHz having an integration time of about 2 μs, the scan would have to be constant over 5–10 picture elements for proper color signal decoding. To maintain NTSC color transmission standards, the scan would have to be held constant to about 3 ns of timing error over the integration time of 5 to 10 picture elements. This is not beyond the realm of practical technology but gives an indication of where design development must be concentrated.

1. *Index Stripe Signal Generation*

In the early index camera systems, exemplified by the NHK unit employed at the 1964 Olympic games, the index signal was generated by the formation of a "black" optical stripe for each set of color group stripes. The duty cycle of the index stripes determines the amount of fundamental component of the index signal generated and has an inverse relation to the amount of color signal generated. In the equal width black, red, green and blue stripe system, the 25% index stripe duty cycle implies at least a 25% light loss for the color stripes.

Thus, another means for generating the index signal is desirable. The single tube Sony index system solves this problem by placing alternate conducting and nonconducting stripes in the transparent conducting layer in contact with the photoconductor (see Fig. 16). In this case, interdigitated conducting stripes create an index signal equal to the color group frequency.

Another means of generating index signals at frequencies higher than that of the color group frequency is exemplified in issued patents by Brandinger (1972) and Pritchard (1973b). These approaches employ conducting and nonconducting stripes in specific patterns to produce the required index signal for decoding the particular system being considered. Generally, the chrominance signal and the index signals are separated by appropriate bandpass filters. The index signal is then divided and multiplied by the proper ratio to produce a carrier at the color group fundamental frequency. This carrier is then used to synchronously detect the color signal components. The reference carrier (based on the index signal) is passed through a limiter to remove the amplitude variations.

The use of striped conductors for generating the index signal has the desirable feature that dark current variations are detectable, that is, the index signal is generated even in the absence of light on the vidicon. Thus, if the limiter threshold point is set appropriately, an index signal is present independent of the scene content. It has been reported (Brandinger, 1972) that the ability of the scanning electron beam to resolve the charge pattern due to the conducting stripes is somewhat greater (by a factor of about 2 to 1) than the ability of the same beam to resolve a charge pattern produced optically (Figs. 22, 27).

Systems that generate an index signal frequency higher than that of the color group frequency have an additional problem of ambiguity in the start-up phase at the beginning of each horizontal scan. Numerous solutions to this problem have been considered and experimentally investigated (Brandinger, 1972). The ambiguity may be determined by using about 3% of the horizontal retrace interval time to scan a fixed index reference area at the extreme edge of the raster at the starting side. A conductive stripe with no photoconductor at this point will produce a large lead stripe signal, in the form of dark current, even with no scene illumination, and provide a pulse that may be processed in turn to start the decoding process in the same phase relationship on every line.

2. Specific High Frequency Index System Description

When a frequency different than the color carrier frequency is chosen for the index carrier an electrical beat between the carriers is generated. Assuming a 4.0 MHz color carrier and a 6.0 MHz index carrier, a 2 MHz beat is generated which falls in the luminance signal band. Similarly, all choices of index carrier frequencies below a 3.0 MHz color carrier are included within the luminance band and produce in-band beats. To overcome these beat problems, it would be desirable to provide a high index frequency. Generally, the optical response of the pickup tube limits the upper available frequency to less than 6 MHz for 1-inch vidicons (as shown in Fig. 22).

An added difficulty with index systems in which the index frequency is above the color group frequency, relates to the "start-up" ambiguity. After each scanned line is completed, the electron beam retraces to the beginning of the next line with an inaccuracy that may be greater than the width of one color stripe. Thus, some means of uniquely identifying the color decoding sequence is required at the beginning of each line. Techniques of optically blanking for a fixed time interval prior to the first stripe, as well as other ambiguity resolving structures, are possible.

A block diagram of a unique single color tube index system is shown in Fig. 23. This system having the index carrier at three times the 4 MHz color carrier does not produce the beats in the luminance band and requires only a single signal lead from the pickup device.

Light from the scene is imaged by an objective lens onto the faceplate. Adjacent to the inner surface of the faceplate are the color encoding filters, a transparent indexing signal structure and a photoconductor. Synchronizing signals are coupled from a synchronizing generator to produce the horizontal and vertical electron beam scanning.

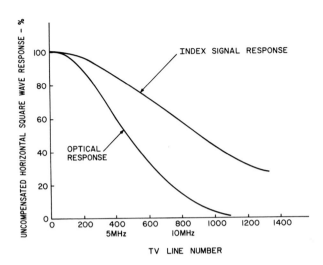

Fig. 22. Typical vidicon optics versus index stripe response.

The color encoding filter and indexing structures are constructed such that when the electron beam scans the photoconductor, a composite electrical signal including the luminance, color carrier, index carrier and ambiguity information is obtained at the signal electrode. In Fig. 24, the frequency spectrum is shown to contain luminance in a 0 to 3.5 MHz band, chrominance extending from 3.5 to 4.5 MHz, and index from 11.5 to 12.5 MHz. The composite signal is preamplified and coupled to two bandpass and one low pass filters. The low pass filter selects the 0 to 3.5 MHz luminance signal. The 4 and 12 MHz bandpass filters select the color and index frequencies. The index signal is limited and frequency divided by a factor of three to generate a 4 MHz reference frequency (the color group frequency). The index signal output from the frequency divider is also coupled to the ambiguity resolving circuitry. This signal is fed in turn to phase shifters to produce three index signal components which decode the color information unambiguously. The color signals are decoded by the three synchronous detectors to reproduce the red, green, and blue picture information. The final combination of luminance and chrominance information in the matrix, followed by an encoder, generates a composite NTSC signal.

The image pickup target is made up of three layers (Fig. 25). The color encoding filter consists of color stripes (dichroic, organic dyes of Fabry–Perot filters). These filters can have the property of transmitting only the primary (or complementary) colors. The color filters are deposited on the substrate by techniques of photoprocessing, printing deposition, or evapor-

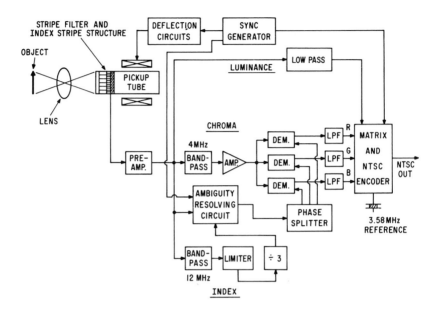

FIG. 23.   Index system block diagram.

ation. The color stripes spatially separate the color information and are arranged parallel to each other and in groups of three. The repetition of these color groups in the direction of the electron beam scan, along with the beam scan velocity, determines the frequency of the electrical color carrier generated. For example, when 200 color groups are used, a

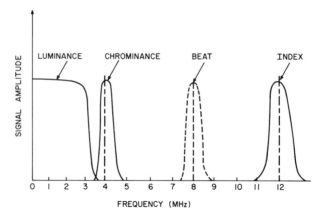

FIG. 24.   High frequency index system signal spectrum.

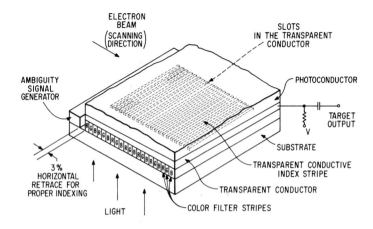

Fig. 25. High frequency index system target structure.

color subcarrier frequency of about 3.6 MHz is generated (with standard NTSC scanning rates). Figure 26 illustrates the angular positioning of the stripes of a color encoding filter with respect to the direction of horizontal scan of an electron beam. The change in angular stripe position permits the electrical frequency generated by the stripes to be varied while maintaining a fixed number of color filter stripes. The frequency versus angle relationship has been shown by R. D. Kell (1956) and is repeated from Section

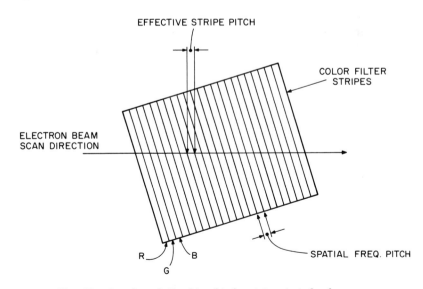

Fig. 26. Angular relationship of index stripes to index frequency.

II, A as

$$f_{\text{electrical}} = f_{\text{spatial}} \, v \, \cos\alpha \qquad (12)$$

where f_{spatial} is in line pairs per millimeter; v is the scanning velocity in millimeters per second, and α is the angle between the stripes and the vertical.

The color stripes are covered by a segmented transparent conductor. These conductors generate the index signal and are made up of stripes (connected outside of the image area) parallel to the color stripes. Parallelism to within 18 mrad, or about 0.01°, is required to maintain NTSC standards of colorimetry. The fabrication of the transparent index stripes may be done by photoresist processing.

The percentage of area covered by the transparent conductor has an effect on the signal output. A duty cycle of 70% results in 70% of the combined RGB color encoding filter stripe area being covered by the transparent conductor and 30% covered by the nonconducting area. As the duty cycle is increased, the total luminance and chrominance signal is increased, while the index signal is decreased. A criterion for choosing a particular duty cycle is that the index signal-to-noise ratio should be sufficient to correctly decode color information in the presence of noise. In Fig. 25, the scale at the bottom of the substrate represents electron beam scanning time in microseconds. In order to travel a full RGB color group, 0.25 μs are required corresponding to a color frequency of 4 MHz. There is one index stripe behind each color stripe which generates 12 MHz when scanned. An 88% duty cycle is illustrated by this example.

The operation of the pickup device is conventional with respect to the photoconductor (Selke, 1969). Light which passes through the transparent substrate is selectively separated by the color stripes and gives rise to a change in the conductivity of the photoconductor. The electron beam charges the photoconductive layer to the cathode potential. The conductive index stripes are positive relative to the cathode. The amount of charge at any point in the photoconductor that leaks to the conductive stripes is proportional to the intensity of light present. The charge deposited by the electron beam during each scan is determined by the amount that leaked off since the previous scan. Hence, the beam current corresponds to the incident light at that particular area and is seen as signal current (i.e., current flow from the transparent electrode, through an external load resistor to the cathode).

When the electron beam scans over the slots in the conductive area with its associated conductive stripes only a dark current signal is generated. Figure 27 is a diagram indicating the direction of the bending of an electron beam as it crosses a slot. The electron beam deposits a negative charge on

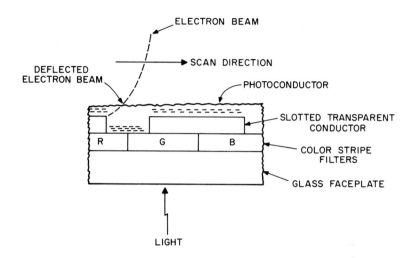

FIG. 27.   Electron beam positioning due to index stripe slot.

the photoconductor overlaying the insulating slots. The magnitude of the negative charge on the photoconductor overlaying the adjacent conductive areas is lessened by the reduced amount of light present. The same is true to a lesser degree for the dark current (no light on the target). Thus, as the electron beam scans over an area of the photoconductor covering a slot the negative charge buildup deflects the beam to the conducting region and increases the beam dwell time on the conducting stripes. The result is an enhanced resolving capability of the pickup tube. Measured improvements of at least 2 to 1 have been obtained relative to the optical resolution. This effect is shown in Fig. 22. Measurements of a standard 1-inch, antimony trisulfide vidicon modified with striped transparent conductive areas with stripe pitches of 500, 1000, and 2000 lpi give useful electrical index signals in the region from 4.5 to 18 MHz. The ability of the tube to resolve this high frequency electrical signal is the basis of an example of a single tube index system (Brandinger, 1972).

The choice of the proper starting point to resolve the ambiguity in color decoding is provided by the signal generated at the break in the transparent conductor as illustrated in Fig. 28a. The uncertainty or jitter in starting time is greatly reduced by the ability of the beam to resolve more than three times the color group frequency. Figure 28b illustrates the timing and generation of the pulse occurring at the start of each horizontal line. The time interval between the conductor-to-dielectric transition and the start of the desired signal output beam completes a single line scan of duration H (horizontal scan time). The time interval T_1 plus T depends on the time required to establish the phase reference. During the interval T_1, the beam lands on the transparent conductor producing a "whiter-than-white" signal current. The time interval $T_1 + T$ includes the equivalent frequency response of the conductor–nonconductor transition, the bandwidth of the electrical circuits used to extract the startup signal, the accuracy of the timing required to establish the startup signal phase, and the signal-to-noise ratio of the startup signal. The ambiguity resolving circuitry is shown in Fig. 29. There are three input signals used to resolve the ambiguity. The first is the composite signal from the pickup tube represented by Fig. 28. This signal information represents the scanning beam's first contact with the transparent conductive layer (time 0). Time T_1 represents the point at which the beam first contacts the substrate. The second input signal is from the index signal frequency divider. The third signal is one representing the start of the horizontal scan period. The control signal is made up of two pulses separated by $1H$. This method illustrates the

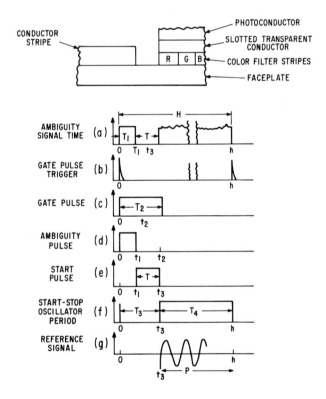

FIG. 28. Ambiguity resolving signal waveforms.

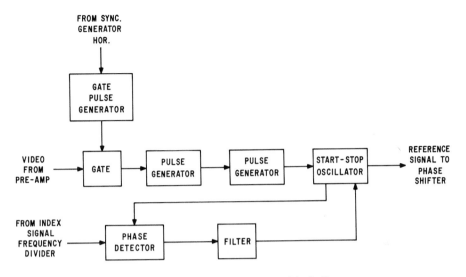

Fig. 29. Ambiguity resolving system block diagram.

principles of only one approach to resolving the decoding ambiguity that has been proved to be practical in experimental setups.

IV. Summary

A number of stripe color encoding single tube, as well as multiple tube, color TV camera systems and devices have been developed and their relative performance characteristics discussed. Examples of several different methods of color encoding were described in detail, including presentation of the fundamental principles involved along with discussion of the operating and fabrication advantages as well as limitations.

A conclusion that may be reached is that, when properly implemented, both frequency division and index systems are capable of producing acceptable color television pictures as characterized by NTSC broadcast performance standards. Each of the systems evaluated has performance advantages and limitations that differ when considered in the light of broadcast, commercial, or consumer applications. No one particular system uniquely satisfies all of the requirements of these applications.

The comparative evaluations described were conducted using a standard antimony trisulfide vidicon. The use of other photosensitive targets employing silicon, lead oxide, or selenide (possibly having a unity gamma exponent) would be expected to provide additional performance gains in the areas of signal-to-noise and intermodulation beat products.

The future development of single tube stripe encoded color camera systems appears to depend largely on tube fabrication technology with commercial application representing the largest current market. Applications in the home consumer market depend not only on low cost cameras, but on the recorders and displays with which the camera will be used. This is currently a specialty, prestige market. The limited quantity, higher cost, high performance color TV broadcast market appears best suited to multiple rather than single tube approaches.

Appendix I. Systems Evaluation Considerations

A. Light and Optics

Many versions of stripe color encoding systems have been applied to single tube pickup sensors. Performance comparisons among these systems require a standardized testing arrangement. A typical experimental setup shown in Fig. 30 employs a standard 8507A 1-inch, separate mesh, antimony trisulfide (Sb_2S_3) vidicon. A relay optical system affords a high degree of flexibility while permitting the use of the same vidicon for each stripe filter system test.

Calibration starts with the characteristics of the light source. A source equivalent to 6500°K (Fig. 31) may consist of a tungsten light source corrected with an IR rejecting filter. Variations of light intensity are ac-

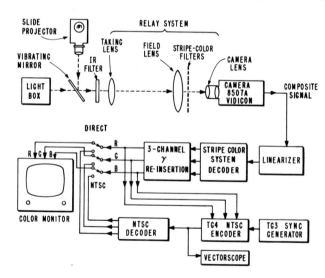

FIG. 30.   Stripe color encoding test setup.

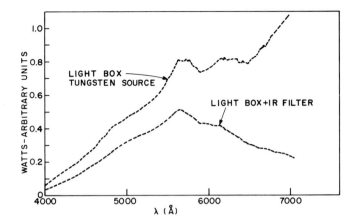

FIG. 31.   Light source spectral characteristic.

complished either by means of an iris or by introducing appropriate neutral density filters.

The relay optical system, consisting of a taking lens, a field lens, and a camera imaging lens, is adjustable over an intermediate image size, at the stripe color filter plane, of from 1.5 × 1.5 inches to almost 3 × 3 inches. Thus, the same pitch stripe color filters may be used over a wide range of spatial frequencies depending upon the intermediate image size chosen. Since the filters may be supplied on individual glass substrates, they can be independently rotated to the desired angular position with respect to the vertical axis of the optical system. Yellow-clear and cyan-clear filters at 100 lpi, and magenta-clear filters at 87.5, 100, and 125 lpi are typical for representative system comparisons. Figures 32 and 33 indicate typical stripe color filter spectral characteristics usable for test purposes. Detailed filter characteristics, given in Appendix III, are appropriate for a variety of amplitude modulation systems.

The experimental circuitry provides encoding and decoding functions for either direct simultaneous viewing of the red, green, and blue signals or standard NTSC format. This allows system evaluation and comparisons to be expressed in terms of commercial broadcast signal standards.

B.   GAMMA

The vidicon (8507A) chosen for evaluation purposes has a nonlinear light input to signal output transfer function with an average gamma exponent of 0.65. The actual exponent varies with light level and has been measured to be in the range of between 0.5 and 0.8. Since the effects of the

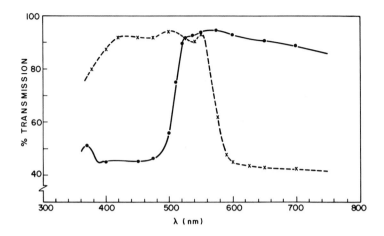

Fig. 32. Stripe color filter spectral characteristics (yellow and cyan).

overall gamma characteristic of a system is an important factor in determining system performance, the following signal processing approach may be taken to best facilitate individual system measurement. The composite signal from the vidicon, after amplification and frequency compensation, is passed through a transfer function "linearizer" unit that corrects the signal to a gamma of unit over a contrast range of at least 10 to 1 (see Fig. 34). The signal may then be decoded and matrixed in linear circuits to produce red, green, and blue output signals in linear form. A three channel gamma reinsertion unit with matched characteristics is then required in the red, green, and blue channels prior to feeding the viewing color monitor in order

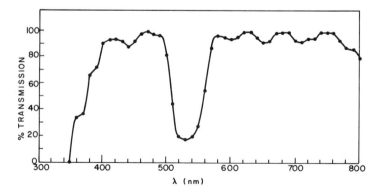

Fig. 33. Stripe color filter spectral characteristics (magenta).

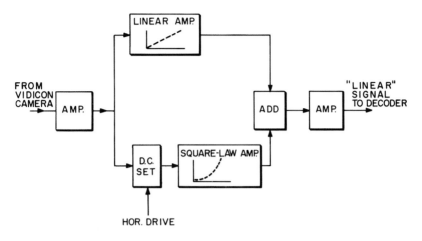

F<small>IG</small>. 34.   Linearizer circuit.

to restore the appropriate overall light-in, light-out transfer function. This experimental approach allows for flexibility in signal processing with a minimum of intermodulation products being generated and with optimum colorimetry performance characteristics.

A typical linearizer circuit uses the square-law characteristics of a field effect transistor (FET) in one channel, operated in parallel with a linear amplifier channel. By combining the outputs of the two channels, the linearizer overall transfer function exponent may be adjusted between the limits of 1.0 and 2.0; the break point can also be adjusted depending upon the point to which picture "black" is set on the FET operating characteristic.

The three channel gamma reinsertion amplifier might typically employ a dc restorer, and a two diode, adjustable break point, nonlinear circuit operating over a 10 to 1 gain ratio. The gamma value is adjusted to approximate a typical value of about 0.5, and the three channels should be matched to within about 1% over the 10 to 1 operating range.

Figure 35 indicates the measured gamma characteristic of a vidicon, as well as the output of the composite signal linearizer prior to the decoding process. The deviation from constant gamma at the highlight end of the range results from the compromise necessary in vidicon beam voltage adjustment to obtain optimum spot focus. The individual systems under consideration treat the final output color signals differently with respect to overall effective transfer function characteristics and are discussed in more detail.

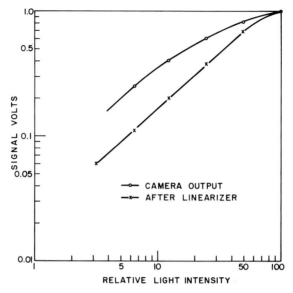

FIG. 35.   Camera and linearizer gamma characteristic.

C. RESOLUTION

Figures 36 and 37 are photographs taken from a line selector oscilloscope indicating the center and corner resolution of the combination of a relay optical system and a typical vidicon camera used for system comparison tests. Figure 38 is a modulation-transfer-function (MTF) curve plotted from this data. The limiting resolution is about 650 TV lines* in the center, and about 575 TV lines in each of the four corners.

Both the camera preamplifier and postamplifier circuits are adjusted to have flat amplitude-versus-frequency response of between 6 and 7 MHz (550 to 600 TV lines) with approximately a 6 dB per octave roll-off. The composite signal linearizer unit is also flat to at least 6 MHz. Thus, the final picture resolution is determined by the limitations of the bandwidths of the luminance bandwidths in the particular color system being evaluated. The overall chrominance channel resolution of 500 kHz at the half-power response point is maintained for all systems.

D. SIGNAL-TO-NOISE

A convenient point at which to measure the signal-to-noise ratio for all systems is at the individual red, green, and blue outputs of the decoder

---

* 80 TV lines = 1 MHz (U.S. Standards).

prior to gamma reinsertion. The following procedure may be used in all cases: The signal to be measured is observed with a line selector oscilloscope to eliminate shading and hum variations. The peak-to-peak signal amplitude is measured from picture black (or from setup, if any) to picture white. The peak-to-peak noise variations are observed by taking the highest value that occurred more than once in a time period of at least one minute. Since the signal at this point is linear, the noise could be measured at any signal pedestal level, but the data is usually taken at about midway in the range between black and white. The signals at this point consist of the individual color low frequency components up to 500 kHz and luminance mixed highs above 500 kHz. An Institute of Radio Engineers (IRE)

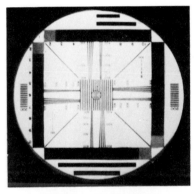

TEST   PATTERN

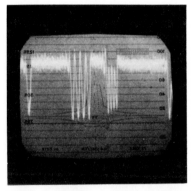

200   LINES - 400   LINES

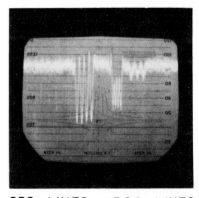

250   LINES - 500   LINES

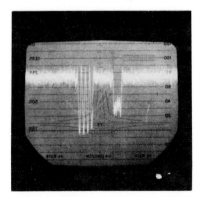

300   LINES - 600   LINES

FIG. 36.   Camera resolution photos—center.

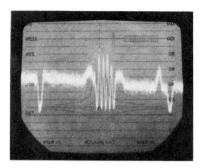

300   LINES

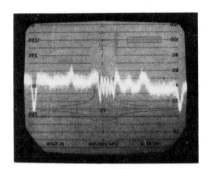

400   LINES

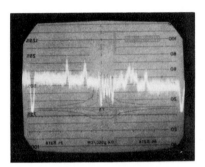

500   LINES

FIG. 37.   Camera resolution photos—corners.

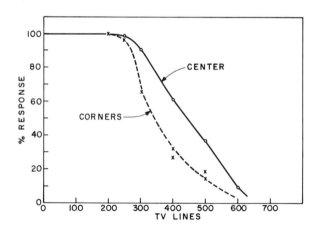

FIG. 38.   Camera MTF curves.

roll-off characteristic filter (20 dB rejection at 3.6 MHz) is always inserted and used as the standard limiting passband characteristic. The peak-to-peak noise is divided by an arbitrary factor of six to convert to rms values. The ratio of peak-to-peak signal to rms noise is computed and expressed in decibels as a measure of the signal-to-noise ratio.

The initial noise characteristic spectrum of a vidicon plus a preamplifier and a postamplifier may be measured with a selective microvoltmeter; a typical set of curves are shown in Fig. 39a and b.

The chrominance information is carried by modulation sideband components of amplitude and/or phase-modulated high frequency subcarriers in all systems. The chrominance information is removed by suitable detection and band-limiting to about 500 kHz. Thus, the high frequency noise in the video spectrum at the subcarrier frequency is converted to relatively low frequency noise and becomes subjectively the major contributor to the overall noise as observed in the picture.

A Percival coil technique may be used in an effort to improve the high frequency signal-to-noise ratio (above the $1/F$ noise range) (Macovski,

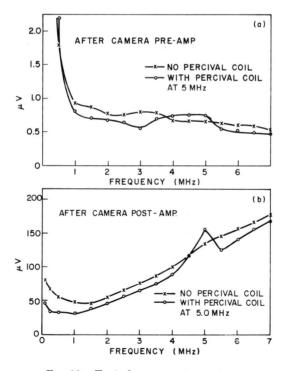

FIG. 39. Typical camera noise spectrum.

1970b). This technique involves a tuned coil in conjunction with the pre-amplifier input capacitance to boost the signal response at a frequency slightly higher than that of the highest subcarrier frequency. Thus, the signal level is increased prior to the addition of noise from the first pre-amplifier input circuit. Care must be taken to subsequently correct the frequency and phase response at a later stage in the preamplifier.

Figure 40 is a diagram of a typical circuit, and the frequency-versus-amplitude curves indicate the degree of increased response at about 5 MHz as well as the degree of compensation obtained in this example. The absolute amount of improvement realizable by this approach depends upon the specific noise source characteristics and upon the amount of incidental circuit feedback introduced by the tuned circuit. Improvements in observable signal-to-noise ratio of about 6 dB are typically obtained by this method.

E. Colorimetry

A basis for colorimetric performance is described by the NTSC broadcast standards* (Federal Communications Commission, 1953). A vectorscope fed from a standard NTSC encoder which, in turn, is fed from the red, green, and blue outputs from the single tube camera decoder, provides colorimetry performance measurements which may be directly translated into terms of hue and saturation as represented by points on the vector-

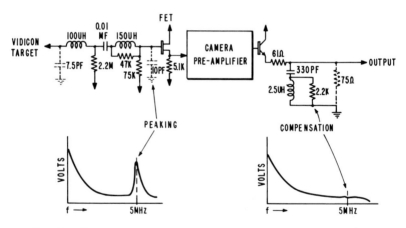

Fig. 40.   Frequency compensation circuit used in camera preamplifier.

---

* FCC broadcast standards are ± 10° phase and ± 20% amplitude for the chrominance subcarrier. The EIA standards of good practice are ± 5° phase and ± 10% amplitude.

scope display (Hazeltine Staff, 1956). Radial distance on the vectorscope is related to saturation (although not a direct measure), while angular position is a direct measure in terms of phase angle of the color hue. The Electronic Industries Association (EIA) chrominance amplitude tolerance of ± 10% and the phase tolerance of ± 5° may be used as the criteria for the colorimetric performance of a stripe filter color camera system.

Optical factors that determine the overall colorimetric performance include the color temperature of the illumination source, the spectral transmission characteristics of the encoding filter stripes, the spectral response of the vidicon, and the spectral response of any trimming filters or antireflective coatings used on any of the optical elements in the system (Jenkins and White, 1957).

Once the source illuminant is chosen and the spectral response of the pickup device and any other trimming filter effects are known, the spectral characteristics of the stripe encoding filters may be calculated. This procedure is discussed in detail in Appendix III. A typical spectral response curve for an antimony trisulfide photoconductor for an 8507A vidicon is shown in Fig. 41.

Small deviations from the ideal spectral characteristic may be compensated for by adjustment of the electronic matrix circuits in the decoder. However, for large deviations, serious colorimetric errors, poorer color signal-to-noise ratio and increased luminance-to-color crosstalk may result due to the increased matrix circuit gain necessary to restore the loss of any particular color sensitivity.

The matrix circuitry should be adjusted to place the vectors representing the red, green, and blue primary colors within the tolerance area on the vectorscope, represented by ± 10% amplitude and ± 5° phase. At the same time, the white balance must be maintained near illuminant C (6500°K). The nature and extent of the colorimetric variations from this starting point may be measured by observing the vectorscope presentation and translating the results into NTSC values.

After the desired stripe filter spectral characteristics are determined, two sources of dynamic operating colorimetric errors exist in systems involving amplitude modulated chrominance subcarriers. The first is a variation in the ability of the scanning electron beam to resolve the stripe color filters equally over the entire raster area. The result is a hue and saturation error in large area color. The variations in spot size and shape result from a complex combination of beam focus and deflection fields. Considerable effort has gone into the development of optimized deflection focusing packages that provide uniform spot resolution of the stripe filters in the equivalent frequency range between 3.5 and 5.0 MHz, and for stripe angles that may vary from the vertical by as much as ± 45°.

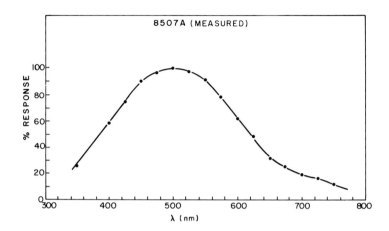

Fɪɢ. 41. Typical vidicon spectral characteristic (measured).

The second source of colorimetric error is in improper tracking of the transfer function characteristic for the red, green, and blue signals over the dynamic operating range from low-lights to high-lights. Although the color subcarrier envelope detectors may be operated in a linear mode, the beam size and shape may change slightly from one subcarrier frequency to the other as the scene content is varied from dark to light. Thus, the effective transfer characteristic can vary for the different colors and produce color errors in mixture colors that vary with scene content. The variation is particularly evident in the case of two carrier stripe color systems where the third color is obtained by mixing the envelope detected signals with the baseband signal. In these systems it is usually necessary to provide individual color channel transfer function correction circuitry to obtain suitable tracking. Three carrier stripe color systems suffer considerably less in this respect, since all three color low frequency components are derived from the detection of modulated carriers which are then matrixed with a common baseband signal. The degree of tracking obtained in four representative systems, without the use of individual channel correction circuits, is indicated in Figs. 42–45.

Systems that rely on a baseband signal and two color subcarriers have an inherent disadvantage in that inequalities in the ability to revolve the color subcarriers result in colorimetric shifts in hue in the direction of the color represented by the information contained within the baseband (usually green). On the other hand, systems that employ three color subcarriers plus a baseband signal have a partial failure mode, when the stripes are inadequately resolved, that is considerably less noticeable to the

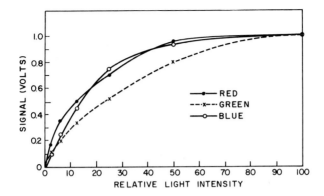

FIG. 42. Gamma tracking characteristic (System I).

human eye. The color error is in saturation rather than hue. Experience indicates that the so-called "fail-to-gray" three stripe color systems are superior in color uniformity and are considerably less sensitive to beam spot nonuniformities.

F. SPATIAL BEATS

A problem that confronts all single tube stripe color encoded camera systems is the optical beat that occurs when the equivalent spatial frequency of the detail content in the scene coincides with the spatial frequency of the color stripes themselves. This pattern manifests itself in the form of low frequency color beats on edges of objects in the scene.

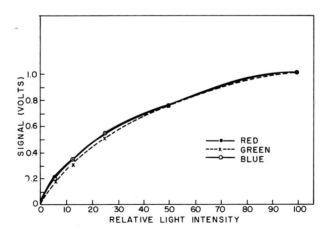

FIG. 43. Gamma tracking characteristic (System II).

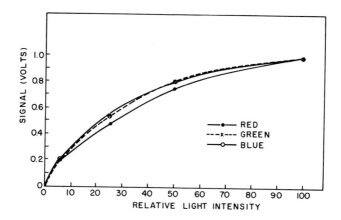

Fig. 44.   Gamma tracking characteristic (System III).

These beats may be essentially eliminated by a variety of optical means for producing a "trap" effect that reduces the scene energy content at a spatial frequency equivalent to the horizontal component of the stripe color encoding filters.

A cylindrical lens that defocuses the scene in the horizontal direction may be used in a fixed-focus situation but is not readily usable for live camera operation. The resulting measured modulation-transfer-function, as shown in Fig. 46, is cosine-shaped and deteriorates the horizontal resolution excessively at frequencies considerably below that of the color subcarrier (Pritchard, 1971, 1973a).

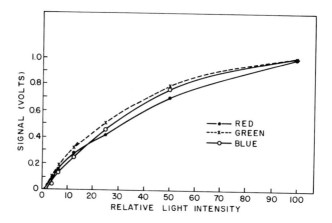

Fig. 45.   Gamma tracking characteristic (System IV).

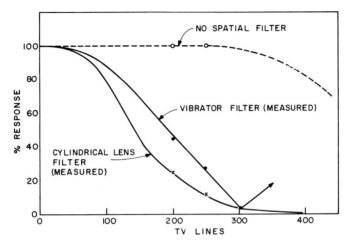

FIG. 46.   Amplitude versus frequency response—spatial filters.

A method that provides a more acceptable modulation-transfer-function characteristic is an amplitude grating oriented in the vertical direction and with a pitch chosen to produce interference pattern cancellation at a spatial frequency equal to that of the horizontal component of the stripe color encoding filter. The light loss, which may be as much as two-to-one, may be tolerated in a film chain camera but would be undesirable in a live pickup. Phase gratings have been used with little light loss, but are generally difficult to construct and require accurate adjustment.

From a performance point of view, the best method employed to date utilizes birefringent materials such as calcite or quartz (Jenkins and White, 1957). These materials have two indices of refraction, and a single light ray is split into orthogonally polarized components called the "ordinary" and "extraordinary" rays. In calcite, the maximum splitting angle is 6°13′, and the two rays exit from the crystal parallel to each other and parallel to the direction of entry into the crystal. Thus, two displaced images are created whose physical separation is dependent upon the splitting angle and the path length through the crystal. The crystal position between the camera lens and the vidicon faceplate is noncritical since the crystal is not an image-forming element. The rotational position is adjusted so that the two images are displaced along the horizontal axis of the scanned raster. When the splitting angle is known, the crystal thickness can be calculated to produce a physical displacement equal to the horizontal component of the width of one color stripe. For calcite, with a splitting angle of 6°, the thickness is approximately 0.010 inch for a stripe color filter whose pitch is 530 lpi and oriented in the vertical direction. The thickness would be increased in proportion to the secant of the angle between the stripes and the vertical. Quartz is used in the same manner, but has a considerably smaller splitting angle (about 0.24°), so that a filter for the same 530 lpi stripes would have thickness in the order of 0.250 inch.

A convenient and flexible method used with an experimental relay optics setup is to produce controlled lateral motion of the image by vibrating the taking lens in the relay optical system with a solenoid fed by 60 Hz alternating current. The degree of image displacement can be adjusted by controlling the current through the solenoid, and thus, the equivalent frequency of the rejection "trap" may be varied to match the system conditions under test. Figure 46 indicates the measured MTF characteristic for such a vibrator system.

Several alternate phase grating filters and their theoretical response characteristics are shown in Fig. 47. The symmetrical sawtooth-shaped filter (A) provides the same response as the birefringent calcite filter. The cylindrical and rectangular grating forms (B and C) result in poor response below 3.5 MHz relative to the A-type grating. Therefore, type A should produce a subjectively sharper picture with better resolution than types

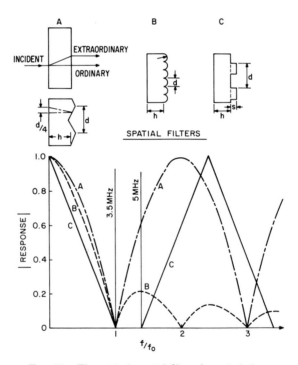

FIG. 47.   Theoretical spatial filter characteristics.

B or C. All of the filter types have a high optical transmission efficiency in the same range as that of the calcite or quartz filters. The phase grating types possess a fundamental cost advantage due to the possibility of using noncrystalline materials. However, contouring of the surface of the substrate materials to the desired shape for phase gratings constitutes a difficult fabrication problem. Successful spatial filters for single tube color cameras have, nevertheless, been reported (Masayuki and Yukio, 1972; Townsend, 1972).

Appendix II. 1-H Delay Comb Filter Fundamentals

A television signal is sampled at repetitive rates of 15.75 kHz, 60 Hz, and 30 Hz resulting from the horizontal and vertical scanning rates. A Fourier analysis of the signal indicates an energy spectrum concentrated at interval spacings of 15.75 kHz, with subsidebands grouped around each 15.75 kHz interval at multiples of 60 and 30 Hz spacing. A second signal, such as the color subcarrier in the standard NTSC format, may have its energy spectrum interleaved with the baseband spectrum by synchronizing the subcarrier with horizontal sync as an odd multiple of one-half line rate (3.579545 MHz in the case of NTSC) (Hazeltine Staff, 1956). In the same manner, two subcarriers may be interlaced with respect to each other by causing them to have a 180° phase relationship with each other when compared on a line-to-line basis.

Two independent pieces of information may be carried as amplitude modulation of the two carriers that are interleaved. At any one instant of time, only one frequency exists, but the independent modulation sideband components may be separated when two successive scanning line periods are compared by means of a comb filter circuit comprised of a direct channel and a 1-H delayed channel (Gunston and Nicholson, 1966; Kallman, 1940; Rakovich, 1970). Figure 48a indicates the basic block diagram of a 1-H delay comb filter, along with vector representations of the operation of the circuit in separating the information contained as amplitude modulation of carriers generated by scanning two color stripe filters, physically oriented so as to produce the required interlaced condition. Figure 48b indicates the two directions of the stripe filters disposed symmetrically with respect to the vertical axis of the scanned raster. Both filters have the same pitch and are set at an angle such that a delay of 90° at the particular carrier frequency generated exists for the red filter, an advance of 90° occurs for the blue filter when compared on a line 1 to line 3 basis. Figure 48c is a vector representation of this situation. In the comb filter circuit, a 90° delay is introduced into the direct path, and the signals

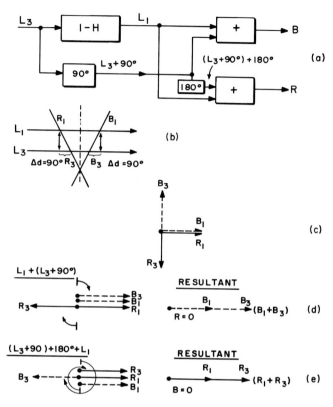

FIG. 48.   Comb filter fundamentals.

are added to the output of a 1-H delay signal in one case and subtracted
(180° inversion) from the 1-H delay signal in the second case. Thus, in one
output the blue information in lines 1 and 3 add while the red information
is canceled. At the same time, the red information adds in the subtractor
circuit while the blue is canceled, as is indicated by the vectors in Fig.
48d and e. The equivalent amplitude-versus-frequency response is as shown
in Fig. 49 for a 1-H delay comb filter using 1-H and 2-H delay elements
(Golding and Garlow, 1971; Rukovich, 1970).

The pitch of the stripe filters along with the standard line scanning rate
determines the absolute value of the subcarrier frequency generated by
the scanning electron beam. If the stripe filters are oriented other than
vertically, the component in the direction of the horizontal scan determines
the electrical frequency generated, while the actual pitch determines the
spatial frequency. The relationship between the spatial frequency, the

electrical frequency generated and the angular displacement for the specific case of ± 90° delay (180° between two symmetrically displaced stripe filters) is shown in Fig. 50.

The 1-H delay device usually employed is a glass acoustic line with ceramic transducers as input and output means. A typical unit may be centered around 3.58 MHz with a bandwidth of about 2 MHz, an insertion loss of 8 dB, matching input and output inductors that result in an impedance of about 560 ohms and a delay of $63.555 \pm .005$ μs.

In practical 1-H delay comb filter circuits, the desired-to-undesired signal ratio is in the range of at least 30 to 40 dB. Since the delay line has a fixed length, practical limits exist as to subcarrier frequency variations within one scan line; i.e., time jitter from one line to the next. Most commercial sync generators have a line-to-line horizontal stability in the order of 3–4 ns and are entirely satisfactory. Time jitter in excess of 8–10 ns causes 7.5 kHz horizontal striations (comb filter imbalance) to become visible.

Scan nonlinearity in the camera causes the frequency of the subcarrier generated by the stripe color filters to vary along the horizontal scan line. The comb filter circuit balance is reasonably insensitive to this type of error, since all frequency sideband components move together as the subcarrier changes value by small amounts (in the order of 1–5%), and if the slope of the phase characteristic in the direct path is made to match that of the delayed path over the range of the expected frequency variation plus sidebands. The equivalent amplitude-versus-frequency response of a

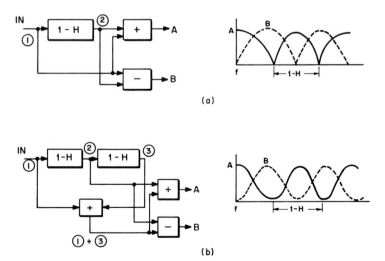

Fig. 49.   Comb filter block diagram and frequency response.

single 1-H delay line comb filter (correlates two scan lines) is in the form of
a full-wave rectified sine wave, as is shown in Fig. 49a. The rejection ratio
is more sensitive to frequency variations than that of a 2-H delay comb
filter (correlates three scan lines) whose characteristic takes the form of a
sine wave, as shown in Fig. 49b. A plot of rejection ratio versus percent
frequency variation, shown in Fig. 51, indicates at least 10 to 1 difference
in balance sensitivity in the practical range of 30 to 40 dB.

In camera systems that use a single 1-H line comb filter to separate two
pieces of color information, the signal contained in two adjacent scanning
lines are correlated. Therefore, the vertical resolution of the *color signal*
is reduced by a factor of 2 to 1 (4 to 2 MHz). The luminance resolution is
unchanged, however. A 2-H delay comb filter reduces the color vertical
resolution by a factor of three, which is still well above the accepted
horizontal color resolution of about 500 kHz.

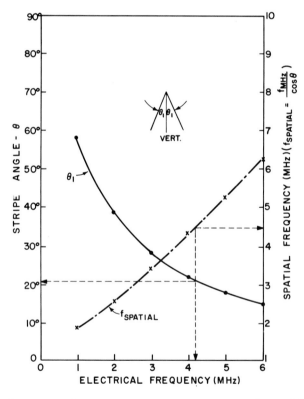

Fɪɢ. 50.   Orientation of stripes versus electrical frequency and spatial frequency.

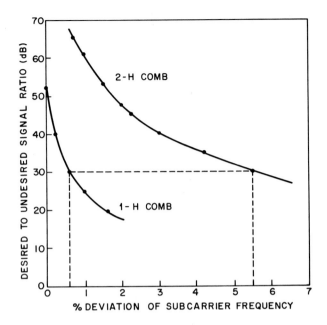

F<small>IG</small>. 51.   Comb filter rejection ratio versus percent subcarrier frequency stability.

The 1-H delay line comb filter technique provides a means for separating two interleaved pieces of color information and thereby effectively allows two color subcarriers and associated modulation sidebands to occupy the same band of frequencies. Therefore, considerable advantages result from the more efficient use of the available bandwidth and may be used to good advantage in developing stripe color camera systems.

Appendix III. Spectral Characteristics of Stripe Filters

An ideal color camera system generates the camera sensitivity curves for blue, green, and red at the respective output terminals in response to unit monochrome light swept through the visible wavelength range. This criteria for camera response leads to a unique determination of the spectral characteristics of the striped filters and the spectral characteristics of the ideal photoconductor target. Since the characteristic of the real photoconductor is unlikely to match the ideal one, a trimming filter placed in the light path compensates for the difference.

The transmission characteristic of two crossed, striped optical filters (Fig. 52) is expressible as the product of two Fourier series:

$$T(x, y, \lambda) = \text{constant} \left\{ \frac{1 - T_c}{2} \left[1 + \sum_0^\infty \frac{4(-1)^n}{(2n+1)\pi} \cos\left\{ \frac{(2n+1)\pi}{2W} \right. \right. \right.$$

$$\left. \left. \times (y \cos \phi - x \sin \phi) \right\} \right] + T_c \right\}$$

$$\times \left\{ \frac{1 - T_y}{2} \left[1 + \sum_0^\infty \frac{4(-1)^n}{(2n+1)\pi} \cos\left\{ \frac{(2n+1)\pi}{2W} \right. \right. \right.$$

$$\left. \left. \times (y \cos \beta - x \sin \beta) \right\} \right] + T_y \right\} \tag{13}$$

$T_c(\lambda)$, $T_y(\lambda)$ are the transmissions of the color selective stripes of the optical filters for light of wavelength λ incident on the photoconductor. Transmission in clear stripes is unity. $4W$ is the spatial period of the optical filters. ϕ, β are the angles of inclination of stripes to the direction of scanning.

The charge pattern $Q(x, y, \lambda)$ on the photoconductor is related to $L(x, y, \lambda)$ by the gamma of the vidicon. If $V(\lambda)$ is the spectral response of the photoconductor,

$$Q(x, y, \lambda) = \text{constant} \, [L(x, y, \lambda) V(\lambda)]^\gamma \tag{14}$$

A simplified form for Q is justifiable if the image of the stripes is sharply defined on the target.

$$Q(x, y, \lambda) \doteq \text{constant} \, V(\lambda)^\gamma \left[\frac{1 + T_c^\gamma}{2} + \frac{1 - T_c^\gamma}{2} \sum_0^\infty \frac{4(-1)^n}{(2n+1)\pi} \right.$$

$$\times \cos\left\{ \frac{(2n+1)\pi}{2W} (y \cos \phi - x \sin \phi) \right\} \right]$$

$$\times \left[\frac{1 + T_y^\gamma}{2} + \frac{1 - T_y^\gamma}{2} \sum_0^\infty \frac{4(-1)^n}{(2n+1)\pi} \right.$$

$$\times \cos\left\{ \frac{(2n+1)\pi}{2W} (y \cos \beta - x \sin \beta) \right\} \right] \tag{15}$$

If either T_y or T_c or both are unity for each wavelength of light in the useful

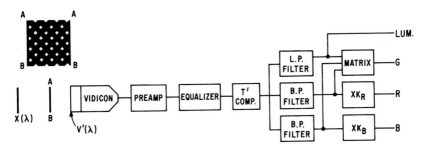

Fɪɢ. 52.   Single tube color camera systems employing two encoding stripe filters.

band, and if only first order terms ($h = 0$) are resolved,

$$Q(x, y, \lambda) \doteq \text{constant } V(\lambda)^{\gamma}\left[\frac{(1 + T_c{}^{\gamma})(1 + T_y{}^{\gamma})}{4}\right.$$

$$+ \frac{(1 - T_c)^{\gamma}(1 + T_y)^{\gamma}}{\pi}\cos\frac{\pi}{2W}\left(y\cos\phi - x\sin\phi\right)\bigg]$$

$$+ \left[\frac{(1 + T_c{}^{\gamma})(1 - T_y{}^{\gamma})}{\pi}\cos\frac{\pi}{2W}\left(y\cos\beta - x\sin\beta\right)\right] \quad (16)$$

A more manageable solution for output signal is obtained if the vidicon spot shape is assumed to be rectangular. The total charge neutralized by a rectangular scanning spot h in height and s in width measured in the direction of horizontal scan is approximated by

$$Q'(x, y, \lambda) \doteq \text{constant}\int_{y}^{y+h}\int_{x}^{x+s}Q(x, y, \lambda)\,dxdy$$

$$= \text{constant } V(\lambda)^{\gamma}\,sh\left[\frac{(1 + T_c{}^{\gamma})(1 + T_y{}^{\gamma})}{4}\right.$$

$$+ F_1(W, \phi, h, s)\frac{1}{\pi}(1 - T_c{}^{\gamma})(1 + T_y{}^{\gamma})$$

$$\times\cos\frac{\pi}{2W}\left(y\cos\phi - x\sin\phi\right)$$

$$+ F_2(W, \beta, h, s)\frac{1}{\pi}(1 + T_c{}^{\gamma})(1 - T_y{}^{\gamma})$$

$$\times \cos \frac{\pi}{2W}\,(y\cos\beta - x\sin\beta)\Bigg] \tag{17}$$

in which

$$F_1(W, \phi, h, s) = \frac{8W^2}{sh\pi^2\,\sin\phi\,\cos\phi}\left\{\left(1 - \cos\!\left(\frac{\pi}{2W}\,s\sin\phi\right)\right.\right.$$

$$- \cos\!\left(\frac{\pi}{2W}\,h\cos\phi\right) + \cos\!\left(\frac{\pi}{2W}\,h\cos\phi\right)$$

$$\times\cos\!\left(\frac{\pi}{2W}\,s\sin\phi\right)\bigg\}^{1/2} \tag{18}$$

$$F_2(W, \beta, h, s) = \frac{8W^2}{sh\pi^2\,\sin\beta\,\cos\beta}\left\{\left(1 - \cos\!\left(\frac{\pi}{2W}\,s\sin\beta\right)\right.\right.$$

$$- \cos\!\left(\frac{\pi}{2W}\,h\cos\beta\right) + \cos\!\left(\frac{\pi}{2W}\,h\cos\beta\right)$$

$$\times\cos\!\left(\frac{\pi}{2W}\,s\sin\beta\right)\bigg\}^{1/2} \tag{19}$$

The vidicon output signal E_1 is proportional to Q'. Setting $(\pi/2W)x\sin\alpha$ equal to $2\pi f_1 t$ and $(\pi/2W)x\sin\beta = 2\pi f_2 t$, where f_1 and f_2 are temporal frequencies there results

$$E(t, \lambda) = \text{constant } V^\gamma sh\left[\frac{(1 + T_c{}^\gamma)(1 + T_y{}^\gamma)}{4} + F_1\frac{(1 - T_c{}^\gamma)(1 + T_y{}^\gamma)}{\pi}\right.$$

$$\times\cos(2\pi f_1 t + \psi_1) + F_2\frac{(1 + T_c{}^\gamma)(1 - T_y{}^\gamma)}{\pi}\cos(2\pi f_2 t + \psi_2)\Bigg] \tag{20}$$

If a third optical filter consisting of alternating magenta and clear stripes is interposed in the light path, it can be shown that the output signal of the vidicon is

$$E(t, \lambda) = \text{constant } V^\gamma sh\left[\frac{(1 + T_c{}^\gamma)(1 + T_y{}^\gamma)(1 + T_m{}^\gamma)}{8}\right.$$

$$+ G_1 \frac{(1 - T_c{}^\gamma)(1 + T_y{}^\gamma)(1 + T_m{}^\gamma)}{2\pi} \cos(2\pi f_1 t + \theta_1)$$

$$+ G_2 \frac{(1 + T_c{}^\gamma)(1 - T_y{}^\gamma)(1 + T_m{}^\gamma)}{2\pi} \cos(2\pi f_2 t + \theta_2)$$

$$+ G_3 \frac{(1 + T_c{}^\gamma)(1 + T_y{}^\gamma)(1 - T_m{}^\gamma)}{2\pi} \cos(2\pi f_3 t + \theta_3) \Bigg] \qquad (21)$$

in which G_1, G_2 and G_3 are functions of spot size and stripe width as in Eqs. (18) and (19). In the following derivations V^γ, $T_c{}^\gamma$, $T_m{}^\gamma$, and $T_y{}^\gamma$ are replaced by V', T_c', T_m' and T_y' and the prime is dropped. In the event that $\gamma \neq 1$, or that gamma correction is not applied to the camera signal, the final expressions for the filter characteristics must be raised to the $1/\gamma$ power.

The dc or luminance component and the three carrier components of $E_2(t, \lambda)$ are separated by low pass and bandpass filtering as shown in Fig. 53 Carrier amplitudes are detected by amplitude detection. Each component may be identified with a range of λ in the visible spectrum by noting that the coefficient $(1 - T_c)(1 + T_m)(1 + T_y)$ has a maximum value in the blue range; $(1 + T_c)(1 - T_m)(1 + T_g)$ in the green range; and $(1 + T_c)(1 + T_m)(1 - T_y)$ in the red range. These detected amplitudes are now set equal to the camera sensitivity characteristics S_B', S_G', and S_R' for blue, green, and red (McRae, 1969).

$$K_B V (1 + T_c)(1 + T_m)(1 - T_y) = S_B'$$

$$K_G V (1 + T_c)(1 - T_m)(1 + T_y) = S_G'$$

$$K_R V (1 - T_c)(1 + T_m)(1 + T_y) = S_R'$$

$$K_L V (1 + T_c)(1 + T_m)(1 + T_y) = S_L' \qquad (22)$$

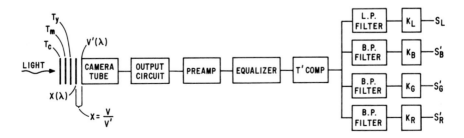

Fig. 53. Single tube color camera systems employing three encoding stripe filters.

in which K_i combine G_i and appropriate gain factors. Solutions for T_c, T_m, and T_y in (11) are

$$T_c = \frac{1 - S_R'K_L/S_L'K_R}{1 + S_R'K_L/S_L'K_R}$$

$$T_m = \frac{1 - S_G'K_L/S_L'K_M}{1 + S_G'K_L/S_L'K_M}$$

$$T_y = \frac{1 - S_B'K_L/S_L'K_B}{1 + S_B'K_L/S_L'K_B} \tag{23}$$

T_c, T_m, and T_y may be determined over certain ranges of wavelength λ merely by inspection of the NTSC camera sensitivity characteristics for positive lobes only (Fig. 54) and are tabulated as follows:

TABLE III

λ	T_c	T_m	T_y	Reason
$\lambda \leq \lambda_{g1}$	1	1		$S_R'(\lambda),\ S_G'(\lambda) = 0$
$\lambda \leq \lambda_{g1}$			0	Blue signal proportional to S_B'
$\lambda_{g1} \leq \lambda \leq \lambda_b$	1			$S_R'(\lambda) = 0$
$\lambda_b \leq \lambda \leq \lambda_r$	1		1	$S_B'(\lambda),\ S_R'(\lambda) = 0$
$\lambda_b \leq \lambda \leq \lambda_r$		0		Green signal proportional to S_G'
$\lambda_r \leq \lambda \leq \lambda_{g2}$			1	$S_B'(\lambda) = 0$
$\lambda \geq \lambda_{g2}$	0			Red signal proportional to S_R'
$\lambda \geq \lambda_{g2}$		1	1	$S_B'(\lambda),\ S_G'(\lambda) = 0$

As a condition that the gain factors K_i be independent of λ, the luminance signal must consist of a linear addition of S_i, namely,

$$S_L' = C_B S_B' + C_G S_G' + C_R S_R' \tag{24}$$

in which C_i are arbitrary. Substituting in (24) for S_L', S_B', S_G', and S_R' from (22), and utilizing Table III, it is found that $K_B = K_L/C_B$, $K_G = K_L/C_G$, $K_R = K_L/C_R$. Since K_L is arbitrary, K_L is set equal to $\frac{1}{8}$ for convenience.

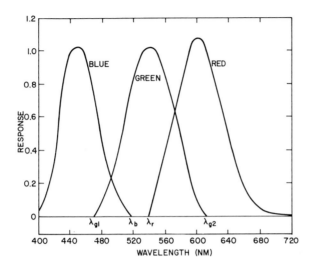

FIG. 54. Camera sensitivities in the standard NTSC system.

Having determined K_i the values of T_i and V in Table III are explicitly given by (23). A complete tabulation of T_i and V over the complete λ range is given in Table IV. Since the spectral characteristic $V(\lambda)$ of the hypothetical camera tube is completely specified in Table IV, a trimming filter $X(\lambda)$ is called for in the real camera system in which the vidicon spectral characteristic may be $V'(\lambda)$ where

$$X(\lambda) = V(\lambda)/V'(\lambda). \tag{25}$$

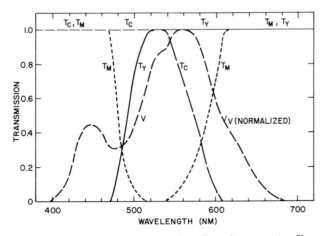

FIG. 55. Spectral characteristics of the encoding filters for two stripe filter systems.

TABLE IV

λ	$T_c(\lambda)$	$T_m(\lambda)$	$T_y(\lambda)$	$V(\lambda)$
$\lambda \leq \lambda_{g1}$	1	1	0	$2C_BS_B'$
$\lambda_{g1} \leq \lambda \leq \lambda_b$	1	$C_BS_B'/(2C_GS_G' + C_BS_B')$	$C_GS_G'/(2C_BS_B' + C_GS_G')$	$\dfrac{(2C_GS_G' + C_BS_B')(2S_B'C_B + C_GS_G')}{C_GS_G' + C_BS_B'}$
$\lambda_b \leq \lambda \leq \lambda_r$	1	0	1	$2C_GS_G'$
$\lambda_r \leq \lambda \leq \lambda_{g2}$	$C_GS_G'/(2C_RS_R' + C_GS_G')$	$C_RS_R'/(2S_G'C_G + C_RS_R')$	1	$\dfrac{(2C_GS_G' + C_RS_R')(2C_RS_R' + C_GS_G')}{C_GS_G' + C_RS_R'}$
$\lambda \geq \lambda_{g2}$	1	1	1	$2C_RS_R'$

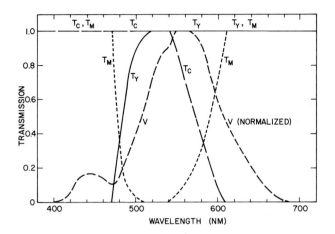

FIG. 56. Spectral characteristics of the encoding filters for three stripe filter systems.

The greatest light economy is achieved when a scale factor is applied to $X(\lambda)$ causing unity transmission at the wavelength of maximum transmission.

The spectral characteristics of the encoding filters T_c, T_m, and T_y and the theoretical camera filter $V(\lambda)$ shown in Figs. 55 and 56 were computed by means of formulas given in Table IV for the NTSC specification of the luminance signal ($0.11\ S_\mathrm{B}' + 0.30\ S_\mathrm{R}' + 0.59\ S{:}'$) and the frequently mentioned combination ($0.25\ S_\mathrm{B}' + 0.25\ S_\mathrm{R}' + 0.50\ S_\mathrm{G}'$). The trimming

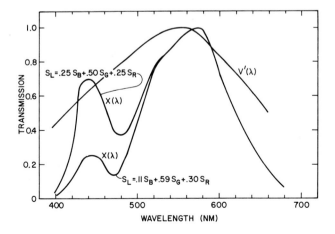

FIG. 57. Trimming filter, $X(\lambda)$.

TABLE V

λ	$T_{\mathrm{c}}(\lambda)$	$T_{\mathrm{y}}(\lambda)$	$V(\lambda)$
$\lambda \leq \lambda_{g1}$	1	0	$2C_{\mathrm{B}}S_{\mathrm{B}}'$
$\lambda_{g1} \leq \lambda \leq \lambda_{b}$	1	$C_{\mathrm{G}}S_{\mathrm{G}}'/(2C_{\mathrm{B}}S_{\mathrm{B}}'+C_{\mathrm{G}}S_{\mathrm{G}}')$	$2C_{\mathrm{B}}S_{\mathrm{B}}'+C_{\mathrm{G}}S_{\mathrm{G}}'$
$\lambda_{b} \geq \lambda \geq \lambda_{r}$	1	1	$2C_{\mathrm{G}}S_{\mathrm{G}}'$
$\lambda_{r} \geq \lambda \geq \lambda_{g2}$	$C_{\mathrm{G}}S_{\mathrm{G}}'/(2C_{\mathrm{R}}S_{\mathrm{R}}'+C_{\mathrm{G}}S_{\mathrm{G}}')$	1	$2C_{\mathrm{R}}S_{\mathrm{R}}'+C_{\mathrm{G}}S_{\mathrm{G}}'$
$\lambda \geq \lambda_{g2}$	0	1	$2C_{\mathrm{R}}S_{\mathrm{R}}'$

filter $X(\lambda)$ appropriate for the compensation of the published spectral characteristics of the RCA vidicon type 8507 is shown in Fig. 57.

A single camera tube system employing only two encoding filters, T_{c} and T_{y} derives the green signal by a matrix operation on the luminance signal and the blue and red signals as specified by (24). Solutions for T_{c}, T_{r}, and V are obtained from (22) and Table III in which T_{m} is set equal to unity for all λ. The results are listed in Table V. It is noted that formulas for T_{c} and T_{y} are unchanged but that the camera spectral response $V(\lambda)$ shown in Fig. 54 differs in the intervals $\lambda_{g1} \leq \lambda \leq \lambda_{b}$ and $\lambda_{r} \geq \lambda \geq \lambda_{g2}$.

The spectral characteristics $T_{\mathrm{c}}(\lambda)$, $T_{\mathrm{y}}(\lambda)$, $V(\lambda)$ and $X(\lambda)$ are shown in Fig. 58. $V(\lambda)$ is the spectral characteristic for the vidicon type from which the trimming filter $X(\lambda)$ is derived according to (25).

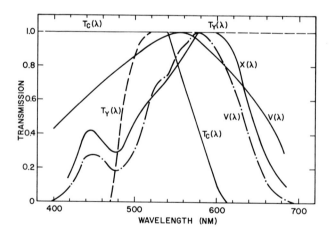

FIG. 58.   Spectral characteristics of encoding filters and field filter.

References

Abrahams, I. C. (1954). *Proc. IRE* **1**, 81.
Borkan, H. (1960). *RCA Rev.* **3**, 3.
Boyd, K. L., and D'Aiuto, J. R. (1973). *Trans. IEEE Broadcast TV Rec.* **19**, 214.
Brandinger, J. J. (1972). U.S. Patent No. 3,735,033.
Brandinger, J. J., Fredendall, G. L., Pritchard, D. H., and Schroeder, A. C. (1974). U.S. Patent No. 3,828,121.
Briel, L. (1970). *J. SMPTE (Soc. Motion Pict. Telev. Eng.)* **4**, 326.
Drummond, J. D. (1969). *Electronics* **42**, No. 14, 114.
Federal Communications Commission (1953). "Compatible Color Order," Sect. 3.682, par. (20)(VII).
Flory, R. E. (1973). *RCA Rev.* **34**, 132.
Fredendall, G. L. (1973). *RCA Rev.* **6**, 267.
Golding, L. S., and Garlow, R. K. (1971). *IEEE Trans. Commun. Technol.* **12**, No. 6, 972.
Goldmark, P. (1970). *J. SMPTE (Soc. Motion Pict. Telev. Eng.)* **8**, 677.
Graham, R., Justice, T. W. H., and Oxenham, M. A. (1961). *IEEE Pap.* No. 3468.
Gray, S., and Weimer, P. K. (1959). *RCA Rev.* **9**, 413.
Gunston, M. A. R., and Nicholson, B. F. (1966). *Marconi Rev.* **29**, 133, 162.
Hayashi, K., and Hayashi, K. (1967). *NHK (Nippon Hoso Kyokai) Lab. Note* No. 113.
Hazeltine Staff (1956). *In* "Principles of Color Television" (M. McIlwain and M. Dean, eds.), pp. 497–500. Wiley, New York.
Jenkins, F. A. and White, H. E. (1957). "Fundamentals of Optics." McGraw-Hill, New York.
Kallman, H. E. (1940). *Proc. IRE* **7**, 302.
Kell, R. D. (1956). U.S. Patent No. 2,733,291.
Konig, S. (1971). *Ferneh- Kino Tech.* **3**, 81, 84.
Kubota, T., and Kurokawa, H. (1971). *Electron Devices Conf., Washington, D.C.*
Kubota, Y. (1972). U.S. Patent No. 3,688,020.
Kurokawa, H., and Kubota, Y. (1972). U.S. Patent No. 3,688,023.
Macovski, A. (1968a). U.S. Patent No. 3,378,634.
Macovski, A. (1968b). U.S. Patent No. 3,419,672.
Macovski, A. (1970a). *J. Appl. Opt.* **8**, 1906.
Macovski, A. (1970b). *IEEE Trans. Broadcast.* **12**, 75.
Macovski, A. (1972). *J. Appl. Opt.* **2**, 416.
McRae, D. H. (1969). *J. SMPTE (Soc. Motion Pict. Telev. Eng.)* **3**, 140.
Masayuki, M., and Yukio, O. (1972). *J. SMPTE (Soc. Motion Pict. Telev. Eng.)* **81**, 282.
Nagahara, S. (1972). *J. Inst. Telev. Eng. Jap.* **2**, 104.
Nobutoki, S., Nagahara, T., and Takagi, T. (1971). *IEEE Trans. Electron Devices* **11**, 1094.
Pritchard, D. H. (1971). U.S. Patent No. 3,588,224.
Pritchard, D. H. (1973a). *RCA Rev.* **6**, 217.
Pritchard, D. H. (1973b). U.S. Patent No. 3,735,028.
RCA (1972). "RCA Vidicon Data Manual CAM-700." RCA Electron. Components, Harrison, New Jersey.
Rakovich, B. D. (1970). *IEEE Trans. Circuit Theory* **2**, No. 1, 41.
Selke, L. A. (1969). *IEEE Trans. Electron Devices* **7**, 625.
Serra, A. (1972). *Rev. Espan. Electron.* **3**, 62.

Sony Corp. (1970). "Service Manual, Color Camera DXC-5000" (Available from Sony Corp.).

Spalding, R. L., Ochs, S. A., and Luedicke, E. (1973). *Proc. IEEE* **10,** 1236.

Takagi, T., and Nagahara, S. (1967). *Jap. Electron. Eng.* **11,** 41, 76.

Takemura, Y. (1971). *Proc. IEEE* **2,** 322.

Takemura, Y., Sato, I., and Tajiri, H. (1973). *J. SMPTE (Soc. Motion Pict. Telev. Eng.)* **1,** 12.

Thompson, R. (1960). U.S. Patent No. 2,962,546.

Townsend, R. L. (1972). *Appl. Opt.* **11,** 2463.

Watanabe, H., Kobayashi, H., Funahaghi, K., and Takemura, Y. (1973). *Toshiba Rev.* **2,** 17.

Weimer, P. K., Gray, S., Beadle, C. W., Borkan, H., Ochs, S. A., and Thompson, H. C. (1960). *IEEE Trans. Electron Devices* **7,** 147.

Yoneyama, M. (1972). *J. Inst. Telev. Eng. Jap.* **26,** 14 and 104.

Zworykin, V. K., and Morton, G. A. (1954). "Television," p. 822. Wiley, New York.

Subject Index

A

Acoustic power, phase shift and, 24
Acoustooptic Bragg diffraction, 23
Acoustooptic deflection
  diffraction angle in, 43
  in laser displays, 40–44
  performance of, 31
Acoustooptic devices
  general concept of, 4
  wavelength-dependent scanning angles in, 5–6
Acoustooptic imaging, with pulsed laser, 55–58
Acoustooptic modulation, 22–27
  for 1125-line color-TV display system, 44–45
  frequency response of, 24, 26
  in laser facsimile system, 52
  laser intensity modulation and, 22
Acoustooptic modulator
  length of, 57
  as spatial modulator, 56
Acoustooptic properties, of various materials, 25
Alphanumeric display devices, 123–137
  interdigital arrays in, 125–128
  longitudinal mode scattering devices and, 123–125
  performance limitations in, 134–137
  PLZT 9/65/35 devices as, 123–137

power considerations in, 131–133
transverse mode interdigital array devices and, 125–137
AO modulation, see Acoustooptic modulation
Argon-krypton gas laser, 12–14

B

Beam deflection
  acoustooptic, 40–44
  digital system in, 41
  galvanometer type, 37–40
  in laser displays, 31–44
  raster irregularity compensation in, 36–37
  resolution capability and scanning speed in, 31
  rotating mirror polygon type, 32–36
Bell Laboratories, 51–52, 120, 159
Birefringence, variable, 68
Birefringence changes
  applied electric fields and, 86–88
  strain biasing and, 85–89
Birefringence devices, transverse mode interdigital-array, 125–136
Birefringence effect, in PLZT ceramics, 67
Birefringence-mode strain-biased image storage and display devices, 83–104
Bivicon camera, 176–178
Bragg diffraction, acoustooptic, 23

247